Biochemistry
Laboratory
Manual

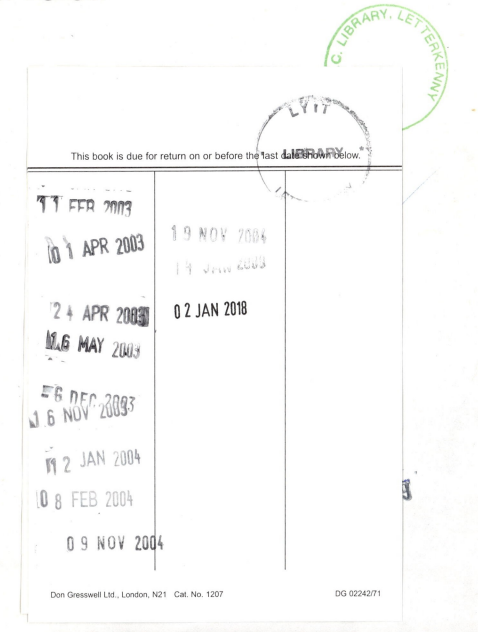

Biochemistry
Laboratory
Manual Third Edition

F.M. Strong
University of Wisconsin, Madison

Gilbert H. Koch
University of Wisconsin, Milwaukee

wcb

Wm. C. Brown Company Publishers

Dubuque, Iowa

Printed in the United States of America
10 9 8 7 6

Contents

Preface

The present manual is the basis for teaching the laboratory portion of a one-semester course in general biochemistry particularly at the freshman and sophomore levels. Such a course for undergraduate students who do not intend to specialize in biochemistry proper is justified in part by the fundamental importance of the subject to many of the aspects and activities of modern life. Perhaps not the least of these is its intrinsic relevance to the natural interest in one's own physical well-being. Of no less importance, however, is its value to students who are planning to enter many professional careers other than biochemistry. For this reason the manual can assist in teaching biochemical techniques and experimental principles to undergraduates majoring in nursing, medical technology, allied health sciences, home economics, nutrition, and biology. The manual has served successfully for pre-professional students in chemistry, medicine, dentistry, and veterinary science as they prepare for a more rigorous biochemical training in these disciplines.

If a student has an adequate high school mathematics background and has taken one or two semesters of college-level chemistry, including some organic chemistry, then the student is fully capable of handling a general biochemistry course at the freshman/sophomore level. The lecture material in such a one-semester course at most universities and colleges consists largely of three parts: (1) biological aspects of water and inorganic elements, ionization of acids and bases, pH, buffers, and a brief review of functional groups of organic compounds; (2) the chemistry of carbohydrates, lipids, proteins, nucleic acids, enzymes and coenzymes; and (3) the digestion, absorption, and metabolism of the major nutrients, energy metabolism, the role of nucleic acids in protein biosynthesis and the transfer of genetic information, and other aspects of modern molecular biology. In conjunction with lecture material of this nature and with occasional explanations by the laboratory instructor, the experimental procedures described in this laboratory manual are well within the capabilities of an undergraduate to perform and understand.

The experiments have been selected to illustrate (a) the most common and useful qualitative tests for major classes of biological materials and their components; (b) as many quantitative methods as seemed feasible; (c) chromatographic and electrophoretic techniques capable of being performed quickly and without the use of complex, expensive instrumentation; and (d) the inclusion of unknowns to stimulate student interest and bring out the value of many of the biochemical techniques and procedures.

Based on the very gratifying results from a survey of our colleagues at other universities and colleges who have adopted the manual for their classes, and on our own experiences with the previous edition, we have made additions, deletions and revisions. Two completely new experiments deal with *Dialysis* and *Colorimetry*. Dialysis is not only an essential physiological process in plants and animals, but it is also an important biochemical technique used primarily in the isolation and purification of macromolecules.. So many biochemical assays involve spectrophotometry, and colorimetry in particular, that we have devoted an entire experiment to set forth the basic principles. The experiment on "Active Acidity" has been divided into two separate and more detailed experiments, namely, *Measurement of pH* and *Buffers. The Control of pH.* Enzymatic methods for the estimation of glucose and cholesterol in blood serum or plasma have been added to update these procedures. However, if the cost of enzymes does not fit your shrinking laboratory budgets, the older methods have been retained. All other procedures have been carefully examined and have been changed or modified for greater clarity and improvement of techniques.

Following the suggestions of reviewers, the experiments in the sections on "Water" and "Inorganic Elements," which were located at different parts in the previous edition have been combined into *Water and Inorganic Elements* and placed at the beginning of the manual. This section includes the new experiments on *Dialysis* and *Colorimetry.* The experiments in this section can now serve as introductory techniques in biochemistry and provide lead-time for the lecturer to develop the material on the major cellular constituents. Based on our survey, we have revised and expanded most of the experiments in the section on *Energy and Metabolism* and we have deleted those which were seldom used.

One of the new features of this edition is a comprehensive and detailed listing in the Appendix of *Reagents, Materials and Equipment* for each experiment. In addition, the quantity of each item is given based on an *enrollment of 100 students,* and these amounts can be altered to fit the size of your class, plus an allowance for waste or spillage. We are aware of the time and effort involved in repeatedly preparing lists for the stockroom, especially calculating the amounts of reagents needed to prepare the numerous solutions. Therefore, we know you will greatly appreciate this listing. At the end of the Procedure for each experiment you will find the exact page number in the Appendix where the reagents and materials are listed. We strongly recommend that *all laboratory instructors read the listing* for a given experiment because *important notations are frequently given.* A recommended supplier may be given for certain reagents or equipment. The compilation of this listing has been a tedious job; but if an inadvertant error has occurred, or if you have a suggestion, please tell us about it.

There are more experiments in the manual than can be used in a one-semester course and, therefore, those experiments can be selected that are the most suitable for the type of student in the class. For example, if the students are majoring in nursing, medical technology, or the allied health sciences, it would be prudent to include experiments on the detection of glucose in urine, the determination of glucose, cholesterol and urea in blood serum or plasma, and the electrophoresis of serum proteins. On the other hand, for home economics and nutrition majors, you can use experiments dealing with the determination of sugars in foods, unsaturation (iodine number) of fats, Soxhlet determination of fat, Kjeldahl determination of nitrogen and crude protein, determination of vitamin C, and the fermentability of sugars. In addition to the above, there are about 28 other experiments which represent the "nuts and bolts" of an undergraduate biochemical laboratory course. This extensive coverage of biochemical experiments makes the manual particularly suitable for a one-semester course devoted exclusively to general biochemistry at the freshman and sophomore level compared to a type of manual covering the three areas of general chemistry, organic chemistry, and biochemistry where of necessity the experiments in biochemistry are very limited in number and scope.

As in the previous edition, all experiments have the same format: *Introduction, Procedure, Laboratory Report,* and *Questions.* The *Introduction* discusses the nature of the experiment, outlines the principles involved, and describes the physical changes or chemical reactions which the biochemical substances studied undergo. Much of this descriptive material has been updated and expanded. The *Procedure* carefully teaches the student how to perform the biochemical techniques so that the experiment can be performed successfully with minimum supervision and maximum comprehension. Where necessary the procedures have been rewritten for greater clarity. The *Laboratory Report* provides space for the recording of data and observations, and for performing calculations in an orderly manner so that the instructor will be able to check these reports readily without tediously trying to decipher the student's intentions. The *Questions* are designed to test the student's comprehension of the introductory and experimental material so that a greater feeling of learning and relevancy has been achieved. Throughout the manual, most of the questions have been changed and many new ones have been added.

Although a distinct effort has been made to keep the required instrumentation to a minimum, every biochemistry laboratory needs certain instruments to function effectively. Among these are ovens, analytical balances, pH meters, photoelectric colorimeters, and possibly a large water bath and a centrifuge. The prominent role of colorimetric methods in current biochemical procedures makes the colorimeter one of the most important laboratory tools. If you cannot afford the expensive equipment for a complete assembly for the Kjeldahl nitrogen procedure or the Soxhlet fat extraction, then a simple demonstration set-up can be used. Few biochemistry faculty appreciate the wide-spread use of these two techniques in the food and brewing industries which are an imporant segment of our national economy.

A few pertinent literature references have been included at the end of most experiments. No effort has been made, however, to document all sources of information, nor is any claim made to originality with respect to any of the experimental procedures.

As is no doubt true of all such compilations, the present work is the end product of an evolutionary process involving the efforts of many individuals. To many generations of students, to a long list of interested and hard working graduate student teaching assistants, and especially to our colleagues, we are deeply indebted for helpful advice, suggestions, and criticisims without which the writing of this manual would not have been possible.

Gilbert H. Koch
F. M. Strong

Introduction

BACKGROUND INFORMATION

Students taking elementary general biochemistry are expected to be familiar with certain information, which they should have received in their high school chemistry and biology courses and in a one- or two-semester college level course in general chemistry. These include the metric system of weight, volume, and length measurements; solubility rules relating to simple inorganic salts; the meaning of the terms mole, chemical equivalents, molarity, and normality; and a reasonable facility in the use of simple arithmetic and an understanding of common logarithms.

LABORATORY TECHNIQUES

In the experimental procedures to be carried out in this course, wherever water is called for without further qualification, distilled water is to be used. A plastic wash bottle included in each student's locker equipment should be kept filled with distilled water. Please avoid unnecesary waste of distilled water.

It is absolutely imperative that chemical reagents used for qualitative and quantitative tests be free from contamination with other chemicals. Good quality chemicals will be provided for the laboratory work. However, such chemicals will immediately become so contaminated as to be worthless (or even worse than worthless since they may cause misleading experimental results) unless the students using such chemicals follow proper techniques to avoid cross-contamination of one reagent with another. This involves care in returning its own stopper to each reagent bottle immediately after use, avoiding any possibility of contact between the stopper and other chemicals which may be, for example, on the working surface of the table or bench where reagents are kept, and the use of whatever precautions may be necessary to avoid spillage. In particular, the student should never remove solids or liquids from a stock reagent bottle by putting into the bottle his own spatula, pipet, or other instrument. This practice sooner or later inevitably introduces extraneous material into the reagent. Similarly, the student must use adequate precautions in pouring liquids from bottles or beakers to avoid the possibility of drops running down the outside of the container. This can easily be avoided if the lip of the vessel from which the liquid is being poured is touched to a stirring rod or to the surface of the new container to remove the last drop of liquid when the pouring is finished.

Quantitative experiments are based primarily on measurements of weight and volume. Special directions for the use of the chemical balance and volumetric glassware are provided. In all quantitative procedures, it is essential to avoid mechanical loss. For example, weighed samples must be transferred to safe containers *at the balance.* Soluble materials being transferred from one container to another must be rinsed quantitatively from the first to the second. Losses from improper use of stirring rods, spatulas, etc., must be avoided. Wherever aliquots of one solution are taken and the remainder of the solution discarded, the volumes involved must be immediately written down so that the necessary calculations may be made later.

The chemical laboratory is potentially a very dangerous place. Many of the chemicals used are toxic or corrosive, while others may be very flammable or even capable of exploding. To avoid accidents, each student must cooperate fully with the instructors in informing himself of potentially hazardous situations and must follow good laboratory techniques. The pipetting of corrosive liquids by mouth should at all times be avoided. Rubber bulbs will be provided for this purpose and should be used. The location and use of fire extinguishers should be pointed out by the instructor and carefully noted by each student. General neatness and good housekeeping are most important safety factors in laboratory work. Likewise, economy in the use of chemicals is important, as many are expensive, and excessive use is not only wasteful but likely to be dangerous.

LABORATORY NOTES AND WRITE-UPS

Probably the most important single aspect of scientific experimentation is the keeping of adequate notes and records. The importance of this cannot be overemphasized. Unless the experimenter is willing to take the necessary time and trouble to make a clear, understandable, and complete record of what was done and what happened when the experimental work was carried out, his time and effort are wasted. Students will be expected to pay a good deal of attention to this subject, and their laboratory grades will depend to a large extent on how well the laboratory records are kept.

The main principles of proper scientific note keeping are as follows: Make the notes at the time the experiment is being done. Enter the record directly into the permanent notebook (this book) and not on loose pieces of paper. The record of each experiment must be dated. All items of data must be clearly labeled so as to show unambiguously what they are or what they mean.

The student's write-up or laboratory report should tell primarily what happened when the experiments were done. Whatever tests are carried out should be observed closely for any changes in color, odor, formation of precipitates, evolution of gases, or any other observable change whatever. The write-up should include an accurate and critical description of such results. Whenever several tests are made, the results should be compared with one another. Tables and graphs should be prepared as needed to present the experimental results in an understandable and systematic manner. Wherever appropriate, conclusions should be drawn as warranted by the experimental evidence. Answers to the questions that are included in all of the experiments should, of course, be given. These answers should be obtainable from the textbook used in the course, from lecture notes, or by consulting the general references given at the back of this manual. If the necessary information cannot be obtained from any of these sources, consult the laboratory instructor.

Students are expected to use reasonably good English in their experimental write-ups and above all to make the record intelligible. Furthermore, the records of the experiments done must be kept reasonably well up to date. As indicated above, a record of all direct observations and quantitative measurements *must* be made at the time the experiment is done. The answers to questions and the complete write-up may be finished after the laboratory period, but in any event must be completed within at least one week of the date on which the experiment was done.

ATTITIUDE TOWARD LABORATORY WORK

The student's attitude and approach to laboratory work are all-important. The mechanical, cookbook type of procedure is worthless. To derive any appreciable benefit from the time and effort spent doing laboratory experiments, the student *must be thinking* about what the experiment is intended to illustrate, what is happening, why various reagents are used, and so forth. You will find the laboratory work much more rewarding if you will approach it in this way, and if you will cooperate in following the other suggestions above.

The Use and Care of Balances

Two types of balances are used in performing the experiments in this manual. (A) General laboratory balances, which can weigh within 10 milligrams (0.01 gram), and (B) analytical balances capable of weighing within 0.1 milligram (0.0001 gram). A general laboratory rule that will be strictly enforced at all times is: *KEEP THE BALANCES CLEAN*. Any spillage must be cleaned up *immediately* after you are through using a balance. Analytical balances and top-loading laboratory balances are *expensive*, precision instruments. Do not abuse them in any way at any time.

A. *General Laboratory Balances. Never* weigh reagents directly on a balance pan. Use the glassine paper provided for solid substances. For corrosive substances, such as NaOH or KOH pellets, use a small glass beaker.

After checking the pan for cleanliness, note the zero point of the balance. If you are weighing a substance *by difference* (where you first weigh a piece of paper or an empty beaker), no adjustment is necessary if the zero point is slightly off. However, if you are weighing an object (*e.g.*, a dish as in Expt. 1—Moisture Determination), which requires reweighing at a later time in the same or next laboratory period, it is important to correct any zero point deviation each time before weighing the object. Your instructor will advise you on the proper procedure.

The triple-beam balance (total capacity, 311 g), or one of its modifications, is probably the most widely used general-purpose, laboratory balance. The front beam has a slider for adjusting the weight between 0 and 1 gram, with major divisions at 0.1 g intervals and smaller divisions of 0.01 g. Be certain that the weights on the two notched beams are always properly seated in the notch, otherwise your weighing will be in error. Extra 100 g weights can be suspended from the end of the beam assembly to increase the balance capacity to 311 g. Quadruple-beam balances have a fourth beam notched for a sliding weight at the 100- and 200-gram positions. Still other models are of the dial-o-matic type, which eliminates beams of the lower weight capacity.

In recent years the top-loading, single-pan balances have been introduced for general laboratory use. Because they are direct, digital-reading, they are more rapid for weighing purposes than the conventional beam-type balance.

B. *Analytical Balances.* The operation of a single-pan analytical balance will not be described here in detail because of the variations in adjustments and weight controls on the different brands of balances. The operational controls and methods of weight readout often vary among the models available from the same manufacturer. For this reason the instructor will demonstrate the operation of the analytical balances that are in your laboratory.

However, the following rules are applicable for the use and care of all types of single-pan analytical balances:

1. Keep the balance compartment *CLEAN*. If necessary, clean the balance pan with a camel hair brush. Remove substances accidentally spilled in the tracks for the sliding doors to prevent impairment of their sliding movement.

2. Use a piece of glassine paper or a small watch glass on the pan when weighing samples or chemicals.

3. *Never* weigh an object while it is still warm. Convection currents of air in the balance compartment will render the weighing meaningless.

4. With the balance in the "Arrest" position, check the balance level. If the balance is not level, adjust one of the feet of the balance.

5. With balance doors closed and with all weights set at zero, put balance in complete release and check the zero point of the balance. Adjust if necessary.

6. Return balance to "Arrest" position. Balance is ready for use.

7. Open balance door to place object on pan. Close door when weighing object.

8. Release the pan and adjust weight knobs as instructed for your particular balance. Balance must be in *complete release* position when making final weighing.

9. Read and record the weight of the object.

10. Return the balance to the "Arrest" position. *The balance must always be in the "Arrest" position when loading or unloading a balance pan.*

11. Set weights back to zero, close balance doors, and clean up any mess *before leaving* the balance.

C. *Practice Weighings.*

(1) Obtain an object from your instructor which weighs between 5 and 50 grams (small beakers, flasks, glass stoppers, etc.). Weigh the object as instructed, record the weight, and have the instructor check your balance reading against your recorded weight.

Name of object: _____Weight: _____ grams

(2) Practice weighing 0.2 to 0.4 grams of sand. (Similar to weighing samples as in Expt. 8.) Weigh a piece of glassine paper and record weight below to four decimal places. Add sand from small beaker using a scoop-type spatula (e.g., Fisher Scientific Co., "Scoopula" or equivalent) until weight has increased by 0.2 to 0.4 grams. *Close balance door.* Record reading below and determine weight of sand.

Weight of sand and paper: _____

Weight of paper: _____

Weight of sand, grams: _____

Return the sand to the beaker and clean up around the balance.

Water
and
Inorganic Elements

Experiment **1**

Occurrence of Water in Biological Materials

INTRODUCTION

Water is one of the most important and abundant constituents of plant and animal tissues. The water content of materials is commonly referred to as the "moisture" content. Most vegetables contain 88-95% moisture; fresh fruits and fruit juices fall in the 82-88% range; and grain seeds and cereals have 8-12% water. The percentages of moisture in materials from animal sources are close to the following values: meats, 40-70%; liver, 70%; blood, 79% and milk, 87%. While vegetable oils and certain processed animal fats (e.g., lard) are essentially moisture-free, such fats as butter and oleomargarines contain 15-16% water. (See References)

The "proximate analysis" of a biological material or food substance includes the percentage of moisture, mineral ash (Expt. 3), fat (Expt. 22), protein (Expt. 27), and carbohydrate by difference or by a method given later (Expt. 15). A knowledge of moisture content is important in food processing and in the preparation of diets. Frequently materials are calculated on a "moisture-free" basis for cost comparisons.

The color change caused by the hydration of anhydrous copper sulfate is useful for the detection of water and the rough estimation of relative amounts in biological materials. For the quantitative determination of moisture, drying of a weighed sample in an oven at 105-110°C is required.

PROCEDURE

A. *Qualitative Test for Water.* (Note: To conserve on laboratory time, the instructor can set up the experiment and the student record his observations of the results.) On separate watch glasses place small amounts of potato, meat, fat, egg white, and a few drops of water as a control. Sprinkle with a very small amount of anhydrous copper sulfate. Allow to stand for at least 15 minutes. Record your observations on the report sheet; note in particular the extent of color change and *relative* amounts of water present in the materials (e.g., large, moderate, little, or none). If the air is humid, cover the samples with inverted watch glasses.

B. *Determination of Moisture.* The moisture content of two samples is to be determined by heating a weighed amount of each in an oven until the water has evaporated, and then cooling and reweighing the dried residue. See the bulletin board for your assignment of one material from each of the following two groups of suggested substances:

Group A   Vegetables (root and leaf), fresh fruits, fruit
juices, milk, lean meat, sausage.

Group B   Seeds, nuts, cereals, bread, flour, powdered
milk, vegetable oils, butter or oleomargarine,
lard, suet or other adipose tissue, hair, leather.

Obtain two aluminum foil dishes and, with a pencil or ball-point pen, inscribe your locker number or your initials on the tab or the bottom of each dish. Weigh one of the two dishes to the nearest 0.01 g, using a triple-beam, dial-type, or top-loading balance. Add approximately 5-6 grams of the assigned substance from Group A to the tared Al dish, again weighing to the nearest 0.01 g. If a liquid sample is assigned, premeasure 5 ml of it in a 25-ml graduated cylinder, and then pour it into the tared Al dish for final weighing. In weighing a liquid or fresh material from Group A in particular, speed in weighing the sample is more important than great accuracy. Why? In a similar manner, weigh the second dish and the sample assigned from Group B. Record all data on the laboratory report sheet at the time the weighings are made.

Place the two dishes and samples into the 110° oven. Slip a 2.5 x 7.5-cm paper containing your name under the two dishes so that you can readily locate your dishes when you are to remove them during the next laboratory period. (**NOTE:** If your class meets only once each week, the instructor will turn off the oven after 24 hours and reheat it several hours before your next class meets.) At the next laboratory period, cool the dishes in a desiccator; the instructor will explain the use and function of the desiccator. Weigh the dish and residue on the same balance you used initially. In tabulating your data, note that the moisture lost during the drying process is the difference between (weight of dish and sample *before* drying) and the (weight of dish and residue *after* drying). It is not necessary to know the weight of the residue.

Calculate the percentage of moisture to *three* significant figures. Show appropriate units. Find out from three other members of the class what results they obtained from their pairs of sample.

REFERENCES

Adams, C. F. 1975. *Nutritive Value of American Foods*, U.S. Dept. of Agr., Handbook No. 456, Washington, D.C.: Supt. of Documents, U.S. Government Printing Office.

Watt, B. K., and Merrill, A. L. 1963. *Composition of Foods*, U.S. Dept. of Agr., Handbook No. 8, Washington, D.C.: Supt. of Documents, U.S. Government Printing Office.

Reagents, Materials, and Equipment are listed on page 270.

LABORATORY REPORT

Name _____ Section _____ Date_____

Experiment 1.   OCCURRENCE OF WATER IN BIOLOGICAL MATERIALS

A. *Qualitative Test for Water*

What color did you observe when anhydrous copper sulfate was sprinkled on the substances listed below? Were the relative amounts of water large, moderate, small, or none?

Potato _____

Meat _____

Fat _____

Egg white _____

Water _____

Type of chemical reaction causing the color change _____

Balanced Equation: _____

B. *Determination of Moisture*

	Group A	Group B
Name of substance used	_____	_____
Weight of dish and sample		
Weight of dish	_____	_____
Weight of sample before drying		
Weight of dish and sample *before* drying (from above)		
Weight of dish and sample *after* drying	_____	_____
Grams of moisture in sample		
Percent moisture in sample		

Percent of moisture in samples analyzed by three other students:

	Group A		Group B	
	Name of sample	% H_2O	Name of sample	% H_2O
1.	_____	_____	_____	_____
2.	_____	_____	_____	_____
3.	_____	_____	_____	_____

Name_____ **Section** _____ **Date** _____

Experiment 1. OCCURRENCE OF WATER IN BIOLOGICAL MATERIALS

1. When anhydrous cupric sulfate reacts with water, which hydrated ion is responsible for the color ob-

   served?   _____

2. Both calcium sulfate and magnesium sulfate can form hydrates with water. Could white anhydrous $CaSO_4$ or $MgSO_4$ be substituted for anhydrous $CuSO_4$ in this experiment? Explain.

3. List at least three sources of error that might enter into the determination of moisture as carried out in part B of this experiment, that is, explain why the % moisture on the same kind of substance may vary. Assume that the oven temperature is correct and exclude obvious errors due to spillage and weighing of samples.

4. Why is a 5-6 gram sample used instead of 1 gram?

5. A baker uses *dried* skim milk that contains 3.0% moisture and costs $1.16 per lb, which on a moisture-free basis would be $1.20 per lb. If liquid skim milk containing 90.5% water costs $1.60 per gallon (8.64 lbs), what would be the comparative cost of 1 lb dry solids from the liquid milk? Show work and units. Ans. $1.95 per lb solids.

INTRODUCTION

The passage of low molecular weight molecules and ions through a semipermeable membrane is called *dialysis*, and water, sugars, amino acids, and simple electrolytes are examples of such substances. Proteins and polysaccharides (e.g., starch) are large, colloidal-size macromolecules that cannot pass through the membrane. Dialysis is involved in the delivery of nutrients from arterial blood to the cells and in the removal of waste products by the venous blood. The purification of blood by the glomerulus in each of the millions of nephrons of the kidneys is another important example of dialysis. The semipermeable glomerular membrane permits the passage of water, blood glucose, amino acids, electrolytes, and waste products, but not the passage of blood proteins and blood cells. Normally, all the glucose, amino acids, and most of the water and electrolytes are reabsorbed into the bloodstream, while the waste products, mainly urea, are excreted in the urine. When the kidneys are diseased and no longer function, the blood is dialyzed and purified by the semipermeable tubing of the "kidney machine."

Dialysis is an important biochemical technique for the isolation and purification of such macromolecules as proteins (including enzymes) and nucleic acids. For this purpose, a seamless, cellulose tubing with an average pore radius of 24Å is commonly used. The wall thickness of the membrane used in this experiment is about 0.025 mm (0.001 inch). The cellulose membrane is manufactured in a flattened condition and it contains glycerine, which acts as a plasticizer, otherwise the dry membrane is too brittle to handle. Traces of sulfur compounds and heavy metal ions are present, and these contaminants are substantially removed by soaking in distilled water in 0.01 N acetic acid, or in EDTA. For use in dialysis of nucleic acids, more rigorous treatments are often used. The wetted dialysis membrane is soft and flexible. It can be sealed at one end by knotting, filled with a solution of the substance to be dialyzed, and closed as a sack by knotting the open end. An air space is allowed when knotting the dialysis tubing after filling to prevent excessive pressure build-up in the bag during dialysis, since water molecules will diffuse inward at a faster rate than outward. The increased pressure can enlarge the pores of the membrane, or can even cause it to burst.

The basic method of dialysis for smaller volumes of 100 ml or less is to suspend the dialysis sack in a larger volume of distilled water. The lower molecular weight molecules and ions diffuse through the semipermeable membrane and will accumulate near the outer surface of the membrane. They will then oppose the process of dialysis as they tend to diffuse back into the dialysis bag. For this reason, the water surrounding the bag is stirred frequently or agitated with a magnetic stirrer. Special rocking and rotating dialyzers are used for larger batches. For complete removal of contaminants from the macromolecules, the external solvent is periodically replaced with fresh solvents or a circulatory dialyzer is used. When purifying proteins and nucleic acids, the external solvent surrounding the dialysis tubing or bag contains a buffer or counter ions to prevent denaturation or other physical changes of the macromolecule. More details on dialysis can be found in the reference given at the end of this experiment.

You will perform the dialysis process by dialyzing milk contained in a cellulose dialysis bag and testing the external dialyzate (diffusate) for some of the components of milk.

PROCEDURE

A. *Dialysis of Milk.* Obtain a 38-cm (15″) length of dialyzing tubing (33-mm flat-width) that has been soaked in distilled water. Tightly tie a double knot about 5 cm (2″) from one end of the tubing. Place a short-stem funnel in the open end and, while firmly holding the tubing onto the funnel stem, slowly pour 50 ml of milk (whole or skim) into the dialysis tubing. Allowing an air space of about 2.5 cm (1″) above the milk, securely tie a double knot at this point. Now fold the dialysis bag into a U-shape and tie together the two ends of the bag of milk. (Use string if necessary.) Rinse the outside of the bag with distilled water if you accidently spilled milk on it. Place the bag of milk into a clean 400-ml beaker and fill the beaker with distilled water until the bag begins to float. Raise the bag every 10-15 minutes and stir the beaker of water.

After one hour, begin the series of tests outlined in part B by measuring the prescribed small quantities of dialyzate (the solution in the beaker) with your 10-ml graduated cylinder, but allowing the dialysis to continue in order to increase the concentrations of the diffused substances still to be tested.

If you do not complete all the tests in part B, transfer about 100 ml of the dialyzate to a labeled 125-ml Erlenmeyer flask and refrigerate the stoppered flask until the next laboratory period. Discard the milk into the sink by puncturing the *underside* of the bag (the milk is under pressure). Do *not* throw the milk into the waste container.

B. *Tests for Dialyzate Components.* Record your observations for each test on the laboratory report sheet and show the tests to your instructor.

(1) *Biuret test for protein.* Set up 3 test tubes as follows: into one tube add 3 ml distilled water (blank), into a second tube add 3 ml of your dialyzate, and into a third tube add 10-20 mg (small amount on tip of your narrow nickel spatula) of casein *plus* 3 ml of water. To *all* tubes add 1 ml of 10% NaOH and then add 8-10 drops of 0.1% $CuSO_4$ solution to each. Compare the colors of the dialyzate and casein tubes with the "blank" tube. A purple-violet color is a positive test for protein.

(2) *Benedict's test for reducing sugars.* To separate test tubes, add 3 ml of water and 3 ml of dialyzate respectively. Add 5 ml of Benedict's reagent to each tube and mix thoroughly. Place them in a boiling water bath for 5-6 minutes. A turbid solution, from which a reddish-orange precipitate of cuprous oxide (Cu_2O) settles upon standing, is a positive test for reducing sugar.

(3) *Chloride ion.* To separate test tubes, add 10 ml of water and 10 ml of dialyzate respectively. Acidify each tube with 20 drops of 3 M HNO_3 and then add 10 drops of 0.5% silver nitrate solution to each tube. A white turbidity due to AgCl is a positive test.

(4) *Phosphate.* Set up 3 test tubes as follows: one contains 5 ml of water, the second tube contains 5 ml of dialyzate, and the third tube contains 5 ml of phosphate control (0.4 mg P/5 ml). To each tube add 25 drops of 3 M HNO_3 acid. (Note: the amount of HNO_3 is important.) Add 5 ml of ammonium molybdate reagent (use graduate) to each solution. Place the three tubes in a water bath heated to 60-70°C. (Do not exceed 70°, otherwise MoO_3 will precipitate.) Occasionally mix each tube while heating. Within 5-10 minutes a yellow precipitate of phosphomolybdate will form in the phosphate control. Remove the tubes and allow them to stand without further heating. Is there much phosphate in your dialyzate? Milk contains appreciable phosphate, but much of it is bound to casein (a milk protein) still inside the dialysis bag.

(5) *Calcium ion.* To 50 ml of dialyzate in a 100-ml beaker, add 3 drops of methyl red indicator and 10 ml of 10% NH_4Cl solution. If the solution is not yellow, add 2 M NH_4OH dropwise with stirring until it turns yellow. Now add 6 M acetic acid dropwise with stirring until the solution turns red. Heat to boiling, add 5 ml of saturated ammonium oxalate solution, and boil for 1 minute. Allow to cool. A white precipitate of calcium oxalate (CaC_2O_4) or turbidity is indicative of the presence of Ca^{++} ions. The CaC_2O_4 precipitate often forms slowly, and you may place the covered beaker in your locker and observe the results at the next laboratory period.

REFERENCE

McPhie, P. 1971. Dialysis. In *Methods in Enzymology,* ed. S. P. Colowick and N. O. Kaplan. vol. 22, pp. 23-32. New York: Academic Press.

Reagents and Materials are listed on page 270.

LABORATORY REPORT

Name_____ Section_____ Date_____

Experiment 2.   DIALYSIS

Describe the results of the tests you performed on the dialyzate. Indicate the color of the test solution in (1), and the color and quantity (slight, moderate, large) of the precipitates in the other tests.

(1) *Protein*

Dialyzate_____

Casein_____

(2) *Reducing sugar*

_____

(3) *Chloride ion*

_____

(4) *Phosphate ion*

Dialyzate_____

Phosphate control_____

(5) *Calcium ion*

_____

Write *net ionic* equations for the reactions of the inorganic ions in (3), (4), and (5) and the test reagents.

(3) Chloride ion and silver nitrate reagent.

(4) Phosphate ion | ammonium molybdate in presence of nitric acid.

(5) Calcium ion + ammonium oxalate.

Name_____ **Section** _____ **Date** _____

Experiment 2.   DIALYSIS

1. Of what material is the dialysis membrane made that is used in this experiment? _____

2. What is the average pore radius and wall thickness of your dialysis tubing?

    Pore size_____   Wall thickness_____

3. Name at least two other materials used for dialysis membranes.

    _____

4. List other waste products in urine besides urea that are removed from blood during dialysis by the kidney. See your text.

    _____

5. Give the name and the molecular weight of milk sugar whose formula is $C_{12}H_{22}O_{11}$.

    Name_____Molecular weight_____

6. In comparison with milk sugar, what is the approximate molecular weight range of proteins? See your text.

    _____

7. Explain why agitation of the solution surrounding the dialysis sack is important to achieve a reasonable amount of dialysis.

Inorganic Elements of Biological Importance. Dry Ashing

INTRODUCTION

The chemical elements required in largest amounts by living organisms are carbon, hydrogen, oxygen, and nitrogen present in the major organic components of tissues: carbohydrates, proteins, lipids, and nucleic acids. Equally essential are the inorganic or mineral elements, so-called because they are more abundant in nonliving material. Sometimes they are also referred to as ash elements, because they tend to form nonvolatile residues when strongly heated. These elements include the metals: sodium, potassium, calcium, magnesium, iron, copper, manganese, zinc, molybdenum, chromium, tin, and cobalt, and the nonmetals: sulfur, phosphorus, chlorine, iodine, and selenium. These are the so-called nutritionally essential mineral elements required by man and higher animals. Some plants are known also to require boron, but do not require sodium, chlorine, iodine, and cobalt.

Analysis of biological material to determine the kind and amount of mineral elements present requires prior destruction of the organic constituents of the sample and conversion of the mineral elements to simple inorganic compounds. This is accomplished by vigorous oxidation, which may be done either in solution with strong chemical oxidizing agents (wet ashing) or simply by heating in air to red heat (approximately 600-700°C) until all carbon is burned away (dry ashing). Dry ashing is essentially a process of burning the organic material to form carbon dioxide and water. Sulfur and phosphorus are oxidized and may form sulfates and phosphates with any metals present. Nitrogen is released mainly as the free N_2 gas. Since metallic and nonmetallic elements are chemically basic and acidic, respectively, they tend to combine during ashing to form salts. If excess metals are present in the sample, they may produce oxides, and these oxides may react with carbon dioxide to form carbonates. Only the carbonates of sodium and potassium will remain in the ash, however, as the carbonates of other metals are unstable at red heat. If the sample contains an excess of nonmetals, these will be largely volatilized and lost. For example, sulfur and phosphorus may volatilize as SO_2 and P_2O_5, respectively. This loss can be prevented by adding, prior to ashing, some basic reagent such as an oxide, hydroxide, or carbonate of sodium, potassium, calcium, or magnesium. Sometimes a peroxide, nitrate or perchlorate, is used as a strong chemical oxidizing agent to hasten the ashing process. **DANGER:** Mixtures of organic matters with strong oxidizing agents may explode violently on heating. Such ashing procedure must be done only with small quantities of material behind a safety screen.

Fruits, vegetables, and dairy products in general contain more metals than nonmetals, while the reverse is true of cereals, meats, and eggs. The so-called acid-base balance of these food materials is, therefore, basic for the first group and acidic for the second. This balance affects the results of ashing as indicated above. Also, during metabolism in the body the first group tends to add to the alkaline materials in the body and the second to acids. Since the animal body must be maintained at a specific pH, approximately neutral, any imbalance in the diet with respect to metallic and nonmetallic mineral elements must be compensated by elimination of excess from the body. This is accomplished in the case of animals via the kidneys and accounts for the wide pH range of urine, which may vary from about four to eight. In solution in body fluids metallic elements may exist as cations (e.g., Na^+, K^+, Ca^{++}) and nonmetals as anions (e.g., SO_4^{--} and HPO_4^{--}), although many more complex combinations also exist in living systems.

Once the biological sample has been properly ashed, the residue is dissolved in dilute nitric acid to give a solution in which the mineral elements of the original sample are present in the form of simple anions and cations. These ions are then qualitatively detected or quantitatively determined by standard methods of inorganic analysis. Table 3.1 on page 14 lists inorganic elements in some typical foods.

PROCEDURE

A. *Preparation of Sample by Dry Ashing.* Place 5 g of dried whole milk or dried skim milk in an evaporating dish and 5 g of oatmeal or other dry cereal in anorther. Set the dishes on wire triangles (not wire gauze), and heat with a low flame under the hood until the materials cease to smoke. Then increase the flame until the bottom of the dish is a dull red, and burn until the charred mass becomes a gray ash. During the burning, break up the charred particles with a blunt glass rod. To do this, remove the hot dish with a metal tongs, place it on a wire gauze on the bench top, and break the lumps. Replace the dish and complete the dry ashing.

Allow the dish to cool, add 10 ml of distilled water and 5 ml of 3 M nitric acid. Note if there is any effervescence due to carbonates in either dish. Heat on a wire gauze *below* the boiling point for 5 minutes, breaking up any large particcples with a blunt glass rod. Filter the solution through a coarse, wet paper into a 50-ml Erlenmeyer flask or beaker. Wash the dish and paper with water until exactly 25 ml of filtrate have been collected. Make the tests on portions of each filtrate as indicated below.

Keep all tests for a discussion with your instructor.

B. *Calcium.* To 5 ml of each filtrate in separately *marked* test tubes, add 1 ml (20 drops) of 10% NH_4Cl and 1 drop methyl red indicator. Add 2 M NH_4OH *dropwise* with constant shaking of test tube until the mixture turns from red to yellow. Make the mixture slightly acid (red color) by adding 4 drops of 1 M acetic acid (**NOTE:** The addition of 2 M NH_4OH to the filtrate from milk causes a precipitate to form that does not completely redissolve by shaking or by the addition of 1 M acetic acid. The remaining turbidity does not interfere with the final test for Ca^{++}.) Heat the test tubes in a boiling water bath for several minutes, add 1 ml (20 drops) of saturated ammonium oxalate to each tube, heat in the boiling bath for 1-2 minutes, remove from the water bath, and allow to stand for 5 minutes. Compare and record the results obtained on the two samples.

Into 2 supported filter funnels, insert 11-cm filter papers and moisten the papers. Place two marked, clean test tubes underneath the funnels. After allowing the calcium precipitates from above to settle, carefully decant all but about 0.5 ml of the liquids from each tube through their respective filter papers, holding back most of the calcium precipitates in their original tubes. Test collected filtrates for magnesium, as directed in C.

C. *Magnesium.* Add 1 ml (20 drops) of 10% disodium phosphate to the filtrates from the calcium precipitation and add 3 ml of 2 M NH_4OH. If a white, crystalline precipitate does not form at once, rub the inside wall of the test tube with a clean glass rod and let stand a half hour. Compare and record results.

D. *Chlorine.* To a second 3-ml portion of each of the *original* ash solutions in clean test tubes, add 3 drops of 0.5% $AgNO_3$. Heat in the boiling water bath for several minutes. Compare and record results.

E. *Sulfur.* To a third 3-ml portion of each of the original ash solutions, add 2 ml of 10% $BaCl_2$ and 1 ml (20 drops) of 3 M HCl. Heat in boiling water bath for several minutes. Compare and record results.

F. *Phosphorus.* To a fourth 3-ml portion of each of the original ash solutions, add 2-3 ml of ammonium molybdate solution. Heat in the boiling water bath until the tubes are just hot to the touch. *Do not overheat.* Let stand a few minutes. Compare and record results.

G. *Iron.* To 10-ml portions of each of the original ash solutions, add 1 ml (20 drops) of 3 M HCl and then add 0.5% potassium permanganate solution drop by drop with mixing until the mixture remains a faint pink. Usually 1-3 drops are sufficient. Finally add 2 ml of 10% potassium thiocyanate and mix thoroughly. Compare the colors and record the results.

Reagents and Materials are listed on page 271.

Name_____ Section _____ Date _____

Experiment 3. INORGANIC ELEMENTS OF BIOLOGICAL IMPORTANCE. DRY ASHING

Tabulate your results and complete the equations given below.

Element	Identifying compound		Relative intensity of test with	
	Formula	Appearance	Dry milk	Cereal
Calcium				
Magnesium				
Chlorine				
Sulfur				
Phosphorus				
Iron				

Write balanced *net ionic* equations for the reactions between the appropriate ionic forms of the elements in the ash filtrate and the test reagents.

Calcium. Calcium ion and ammonium oxalate reagent.

Magnesium. Magnesium ion and disodium hydrogen phosphate reagent in presence of ammonia.

Chlorine. Chloride ion and silver nitrate reagent.

Sulfur. Sulfate ion and barium chloride reagent.

Phosphorus. Phosphoric acid with ammonium molybdate reagent in presence of nitric acid.

Iron. Ferric ion with potassium thiocyanate reagent.

Table 3.1
Inorganic Elements in Some Typical Foods

Substance	% H_2O	% Ash	Milligrams per 100 g edible portion					
			Ca	Mg	P	Fe	Na	K
Milk, dry whole	2.0	5.9	909	98	708	0.5	405	1330
Milk, dry skim	3.0	8.6	1308	143	1016	0.6	532	1745
Milk, fluid skim	90.5	0.7	121	14	95	trace	52	145
Oatmeal, dry form	8.3	1.9	53	144	405	4.5	2	352
Banana	75.7	0.8	8	33	26	0.7	1	370
Carrots	88.2	0.8	37	23	36	0.7	47	341
Orange juice	88.3	0.4	11	11	17	0.2	1	200
Beef, good grade	60.3	0.8	11	21	171	2.8	65	355
Liver, beef	69.7	1.3	8	13	352	6.5	136	28

Data from B. K. Watt and A. L. Merrill, *Composition of Foods,* U.S. Dept. of Agr., Handbook No. 8, Washington, D.C. Supt. of Documents, U.S. Government Printing Office, 1963.

QUESTIONS

Name_____ Section_____ Date_____

Experiment 3. INORGANIC ELEMENTS OF BIOLOGICAL IMPORTANCE. DRY ASHING

1. List the symbols of the chemical elements needed for life of plants and animals.

   Metals _____

   Nonmetals_____

2. Which of the above elements are *not* required

   by animals?_____ by plants?_____

3. What is the chief type of chemical change that occurs during ashing?

   _____

4. In general, how do the compounds in the ash differ from the original form of the elements in the plant or animal? Illustrate with a few specific examples of compounds from phospholipids, proteins, nucleic acids, etc.

5. Explain the effervescence (or lack of it) when HNO_3 was added to the ash of

   dry milk._____

   the cereal._____

6. What is meant by the acid-base balance of foods?

7. How does the acid-base balance affect the loss of certain elements during dry ashing?

8. a. Would the eating of cereal (*e.g.*, oatmeal) without added milk tend to increase or decrease the acidity of certain body fluids?

   b. What effect would the addition of milk to cereal have on the acid-base balance of this combination? Explain.

   c. If there is an imbalance between metals (cations) and nonmetals (anions) in a diet, how does the body attempt to achieve an acid-base balance near neutrality?

9. The residue from the dry ashing of 5.00 g of dry skim milk weighed 0.395 g and from 5.00 g of oatmeal the ash was 0.095 g. Calculate the percent of ash in each case.

Basic Principles of Colorimetry. The Phenanthroline Method for Iron

INTRODUCTION

The principles and techniques of colorimetry are introduced at this point because colorimetric methods are widely used in the determination of many of the biochemically-important mineral elements as well as numerous biochemical compounds in plant and animal tissues. You will have the opportunity to study the quantitative estimation of phosphorus and iron (Expt. 5), of glucose in blood serum or plasma (Expt. 14), of cholesterol (Expt. 19), and of protein (Expt. 28).

Sometimes a colored constituent of biological tissues, or of a food, can be extracted and colorimetrically determined directly. In many instances, however, the constituent is transformed into a highly colored substance by the reaction with a specific reagent, and the intensity of the colored solution from such a reaction is related to the concentration of the desired constituent. The reaction of iron(II) with 1,10-phenanthroline is a typical example. In still other methods, the constituent from a biological material may initiate a series of reactions, and the intensity of the color of the final product is proportional to the concentration of the original constituent.

Let us now consider some of the basic principles of colorimetry. Under appropriate conditions, the transmission of monochromatic light (i.e., light of a single wavelength) through a homogeneous, transparent solution follows two important laws: *Lambert's law* and *Beer's law*. If a beam of this incident light enters a solution containing molecules of ions having chromophoric (light-absorbing) properties, a fraction of the incident light will be absorbed by these molecules or ions, and the remainder of the light emerges from the solution as transmitted light. With certain factors held constant (e.g., wavelength, solvent, temperature, concentration of absorbing substances, etc.), *Lambert's law* states that the amount of incident light that is absorbed is directly proportional to the thickness or length of the absorbing solution through which the light passes. That is, each unit layer of the solution will absorb the same fraction of incident light. For example, if a chromophoric substance in solution absorbs 10% of the incident light passing through a 1-cm layer or thickness of this solution, then it absorbs 20% of the incident light when passing through a 2-cm layer of solution, and 30% when pasing through a 3-cm layer, etc.

According to *Beer's law*, the fraction of incident light absorbed is directly proportional to the *concentration* of the absorbing molecules or ions, provided the thickness of the solution (the length of the light path) and the factors mentioned earlier, except concentration, are held constant.

The two laws have been combined into the *Lambert-Beer law*, which can be expressed mathematically as follows:

$$(1) \quad \log \frac{I_o}{I} = \epsilon c l \quad \text{or,} \quad (2) \quad A = \epsilon c l$$

where $I_o =$ intensity of the incident light
$I =$ intensity of the transmitted light
$c =$ concentration of light-absorbing constituent
$l =$ length of the light path (in cm) or thickness of the absorbing solution
$\epsilon =$ extinction coefficient of absorbing constituent
$A =$ absorbance, where A in equation (2) replaces the term, $\log I_o/I$, of equation (1). Currently, absorbance (A) replaces the older term, optical density, O.D.

In actual practice, neither the intensity of the incident light, I_0, nor the intensity of the transmitted light, I, is individually measured. Rather, we adjust the instrument meter to read 100% transmittance when the incident light passes through the pure solvent or a "blank" (pure solvent plus reagents used, excluding the constituent being determined). We then insert the sample solution (or standard) to determine the *percent* of incident light transmitted through the sample solution (or standard). The ratio of I/I_0 is called the *transmittance* (T), and the *percent transmittance* (%T) is $100 \times T$.

As indicated by equations (1) and (2), the absorbance, A, is equivalent to $\log_{10} I_0/I$. From this equivalency, the relationship between absorbance and percent transmittance can be developed as follows:

$$(3) \quad A = \log_{10} \frac{I_0}{I} = -\log \frac{I}{I_0} = -\log T = -\log \frac{\%T}{100} = 2 - \log \%T$$

Therefore, the absorbance can be calculated from the percent transmittance by the equation:

$$A = 2 - \log \%T$$

The meter of a spectrophotometer shows a double scale of A and %T.

A spectrophotometer is an instrument with an optical system that produces monochromatic light whose wavelength is expressed in terms of nanometers, abbreviated nm, where $1 \text{ nm} = 1 \times 10^{-9}$ meters. For colorimetry, the visible spectrum falls in the 400-700 nm range. Generally, the light is not absolutely monochromatic, but consists of a narrow band of the light spectrum. The spectrophotometer also has a holder for a test tube (cuvette) for the "blank" or the sample solution. While a large variety of good instruments are available, many of the principles of operation are similar. They may differ in the use of a grating or a prism to produce monochromatic light, and whether the light path is a "single-beam" or a "double-beam." Figure 4.1 shows the controls and the optical system of the Spectronic 20, a single-beam spectrophotometer. Some of the more expensive and sophisticated instruments permit the use of ultraviolet (UV) light for the detection and determination of molecules, which absorb in the UV range of 200-400 nm or in the near infrared (near IR) range of 700-900 nm. The heterocyclic bases of nucleic acids, certain amino acid groupings in proteins, and some vitamins, such as riboflavin, are a few examples of ultraviolet light-absorbing molecules.

In colorimetry, every molecule or ion that has the capability of absorbing light in the visible spectrum will do so at some particular wavelength. For example, if ordinary white light passes through a solution that appears yellow to the eye, the solution is yellow because a chromophoric substance in the solution has absorbed the blue light of the visible spectrum. It is necessary in colorimetry to select that particular wavelength of light at which the chromophoric molecule or ion being determined has the *maximum absorption*. By using this particular wavelength of monochromatic light, we are able to eliminate or reduce the interference of other chromophoric substances in that same solution, and we can now determine the concentration of the desired constituent according to Beer's law; that is, the absorbance of the desired substance at the selected wavelength is directly proportional to its concentration.

The Lambert-Beer law, $A = \epsilon c l$, is obeyed by solutions of simple colored substances, provided we strictly adhere to the required conditions. However, this is not always possible. For example, the incident light may not be absolutely monochromatic, or there may be interfering colored substances. When the absorbances of a colored constituent are too high (i.e., above 0.9), the plot of absorbances against concentrations may deviate from the straight-line relationship of Beer's law. For greater accuracy, the absorbance readings should be in the range of 0.1 to 0.9 (72%T to 12%T). Should the absorbance of the developed color of a sample solution be too high, do not dilute it with water to lower the reading. Serious errors may result due to an equilibrium shift, association, or dissociation of colored complexes. A new sample of proper concentration should be prepared.

Some of the deviations from Beer's law can be minimized by preparing a "standard curve" with the pure compound or ion being analyzed, and comparing the absorbance of the colored substance developed from the unknown sample with the standard curve to obtain the concentration of the desired constituent.

In this experiment the iron(II)-phenanthroline complex ion is used because the color is very stable, and the student need not hasten to complete the experiment as would be the case with some other color-producing reactions. (Note: The absorbance of the color produced in the determination of phosphate in

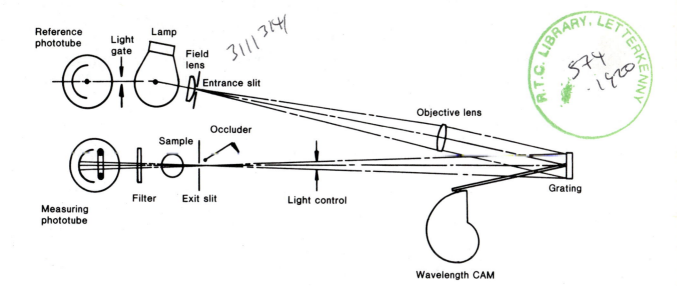

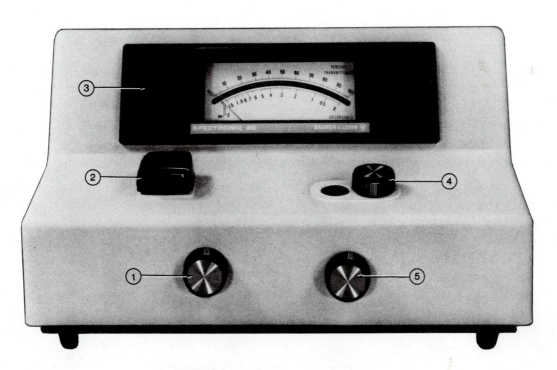

1. Power switch/zero control   3. Pilot lamp
2. Sample holder   4. Wavelength control
5. 100%T control

Figure 4.1. The optical system and a front view of the Spectronic 20. (Courtesy of Bausch and Lomb)

Expt. 5 must be read within 15-30 minutes because the color fades.) The iron must be completely in the ferrous (Fe^{++}) form in order to form the orange-red iron(II)-phenanthroline complex; hydroxylamine reduces any ferric iron to the ferrous state. The complex will form over a wide pH range of 2 to 9, but forms most rapidly at about pH 3.5 and, in addition, at this lower pH the precipitation of certain iron salts is prevented.

The selection of the wavelength of light that gives the *maximum absorbance* will be determined for the iron(II)-phenanthroline complex. Using the wavelength of maximum absorption, a standard curve will be developed with a series of iron samples of increasing concentrations. Finally, a sample of unknown iron concentration will test your technique in colorimetry.

PROCEDURE

A. *Preparation of Standards and Unknown Sample.* You are working with iron concentrations at the microgram per ml level. Carefully clean 8 test tubes (150-mm length) and rinse them thoroughly with distilled water. Allow them to drain completely; they can be oven-dried, or rinsed with acetone and air-dried. Mark the tubes on the frosted area as follows: B (for blank), numbers 1 through 5, and U-1 and U-2 (for duplicate unknowns). Set up the tubes according to the steps given below. It is important that all volumes be measured accurately. With the exception of the unknown sample, all solutions of reagents are available in 25-ml burets on the reagent bench.

1. Pipet 1.00 ml of your unknown into tubes U-1 and U-2. Your instructor will show you the proper pipetting techniques. *Do NOT fill the pipet by mouth suction.*

2. To tubes 1 through 5, respectively, add exactly 1.50, 3.00, 4.50, 6.00, and 7.50 ml of the standard iron solution (1 ml = 10 micrograms Fe). Use the buret provided.

3. Add exactly 1.00 ml of hydroxylamine hydrochloride ($NH_2OH \cdot HCl$) solution to each of the 8 tubes, including the "blank." *Mix.*

4. Add exactly 1.00 ml 1.5 M sodium acetate solution to each tube.

5. Add exactly 2.00 ml of 0.1% phenanthroline solution to each tube. Mix.

6. Add distilled water to each tube to give a final volume of 15 ml as follows:

Tube	B	1	2	3	4	5	U-1	U-2
ml H_2O	11.0	9.5	8.0	6.5	5.0	3.5	10.0	10.0

7. Cover each tube with a piece of Parafilm, or with a *clean* rubber stopper. Mix the contents of each by 5 or 6 complete inversions to ensure homogeneous solutions. Complete color formation of the iron-phenanthroline complex occurs within 5-10 minutes.

B. *Determination of Wavelength at Maximum Absorbance.* Study figure 4.1 to become familiar with the controls and optical system of the Spectronic 20. Consult the "Notes on the Use of the Spectronic 20" given below, and follow the instructor's suggestions after the demonstration of the use of the instrument.

You will make a spectral scan of the iron-phenanthroline in the visible color region of the spectrum from 400 nm to 700 nm. Obtain a pair of matched 12-mm (1/2″) cuvettes. Set the wavelength of the spectrophotometer at 400 nm and adjust the meter needle to 0%T in accordance with Note 2 below. Fill one of your cuvettes with the "blank" solution, and adjust the meter needle to 100%T (zero absorbance) according to Notes 3 and 4 below, and repeat procedure as in Note 5. Now fill the second cuvette with the solution from standard tube No. 5 and take the absorbance and %T readings as explained in Notes 6 and 7. Record your data on the report sheet.

Set the wavelength at 450 nm and repeat *all* steps (Notes 2 through 7). **NOTE:** *Whenever the wavelength setting is changed, the instrument must be readjusted for 0%T and 100%T before reading the absorbance of the sample.* All absorbance readings in this part of the experiment at the various wavelengths of light will be made *only* on standard solution No. 5.

Repeat the above procedure at 475, 500, 510, 525, 550, 600, and 700 nm. Absorbance readings of the sample are made at closer wavelength settings in the region of *maximum* absorbance. According to the literature, this wavelength for the iron(II)-phenanthroline complex is 508 nm.

Return standard No. 5 solution to its original test tube and rinse the cuvette. The "blank" cuvette can be used in part C.

C. *Preparation of a Standard Curve and Determination of Fe in an Unknown.* Determine the absorbance readings for the five standard solutions and the duplicate unknown samples prepared in part A. All absorbance readings are made at the wavelength setting of 508 nm. Check this setting on the spectrophotometer before you start. Follow the notes given below. Record both the absorbance and %T readings. If you incorrectly read the absorbance scale, you can recheck your reading using the %T, or calculate the absorbance from $A = 2 - \log \%T$. Record all readings on the report sheet so you don't lose them. Prepare a standard curve and determine your unknown Fe as μg Fe per 1 ml sample used in part A.

Notes on the use of the Spectronic 20

1. Set the instrument at the desired wavelength.
2. Adjust the instrument needle to read 0% on the Transmittance scale by turning the lower left-hand knob. Be sure cover of cuvette holder is closed.
3. Rinse a colorimeter tube (cuvette) with "Blank" solution and then fill about two-thirds its capacity. Wipe and polish the lower half of the cuvette with a tissue. Keep your fingers off the lower part of the tube. Raise the cover and insert the cuvette into the cuvette holder as far down as it will go. The vertically-etched line at the top of the cuvette should line up with the indicator line at the top of the plastic cuvette holder. Close the cover so that outside light does not affect the phototube.
4. Adjust the needle to read 100% Transmittance by turning the lower right-hand knob.
5. Remove cuvette containing "Blank" and repeat step 2, readjusting lower left-hand knob if necessary. Reinsert "Blank" cuvette and repeat step 4, readjusting right-hand knob if necessary.
6. Rinse, and then fill a second cuvette with the first standard solution. Record the Absorbance reading, estimating to 3 decimal places. Continue with other samples.
7. Check the settings of the instrument frequently to assure yourself that the needle has not drifted from your original settings made in steps 2 and 4.

REFERENCE

FRANSON, M. A., ed. 1976. *Standard methods for examination of water and wastewater*. Washington, D.C.: Amer. Public Health Assoc. Publications.

Reagents, Materials, and Equipment are listed on page 271.

Name_____ Section_____ Date_____

Experiment 4. BASIC PRINCIPLES OF COLORIMETRY. THE PHENANTHROLINE METHOD FOR IRON

B. *Determination of Wavelength at Maximum Absorption.*

Record the Absorbance and the % Transmittance for each wavelength setting for standard No. 5.

Wavelength nm	Absorbance A	Transmittance %T	Wavelength nm	Absorbance A	Transmittance %T
400			525		
450			550		
475			600		
500			650		
510			700		

On the graph paper, plot absorbance (A) on the vertical axis (ordinate) and the wavelength (nm) on the horizontal axis (abcissa). Starting with the origin of the graph, label the vertical axis so that $1'' = 0.100$ absorbance units, and the horizontal axis so that $1'' = 50$ nanometers (nm) with 400 nm on the origin.

Maximum absorbance is at_____nm.

C. *Preparation of Standard Curve and Determination of Unknown.* Record the absorbance and % transmittance of each of the standards and the duplicate unknowns. Calculate the micrograms of iron in each of the standard tubes, where the concentration of the standard Fe solution is 10 μg Fe/ml. The amount of Fe in your duplicate unknown tubes will be determined from your standard curve.

Unknown Sample No. _____

	Standard Fe Tubes					Unknown	
Tube No.	1	2	3	4	5	U-1	U-2
Ml Fe std./tube	1.50	3.00	4.50	6.00	7.50	----	----
μg Fe/tube							
Absorbance							
% Transmittance							

Standard Curve. Prepare a standard curve by plotting the absorbance on the ordinate and the micrograms of Fe per tube on the abcissa. Start the abcissa scale at 10 μg Fe at origin, using the shorter edge of the graph paper, and let $1'' = 10$ μg Fe.

From the standard curve, determine the μg Fe in the unknown tubes. These values represent the Fe in the original 1 ml of unknown given to you by your instructor.

QUESTIONS

Name_____ Section _____ Date _____

Experiment 4. BASIC PRINCIPLES OF COLORIMETRY. THE PHENANTHROLINE METHOD FOR IRON

1. What is meant by the following terms?

Monochromatic light. _____

Chromophoric group or substance. _____

_____

2. The factors that affect the amount of absorption of incident light by a solution of a chromophoric substance include: wavelength of the light, concentration of the chromophoric substance, temperature of
the solution, length or thickness of the solution, solvent used.
Which one of these factors is involved in

Lambert's law? _____

Beer's law?_____

3. If 20% of the incident light is absorbed when passing through a solution that is 10 mm thick, what % of
the incident light is absorbed by this solution if its thickness (length of light path) is

a. 1 millimeter? _____ %   b. 2 centimeters   _____ %

c. Which of the two laws is followed? _____

4. In this experiment, and others in the manual, where the concentration of a chromophoric substance is
determined, which law, Lambert's or Beer's, is the important one? Why?

_____

_____

5. What is the approximate wavelength range in nanometers of each of the following portions of the
spectrum indicated?

Visible_____nm   UV_____nm   Near IR _____nm

6. List 3 examples of biochemical substances that strongly absorb ultraviolet light.

_____

7. Between what limits on the absorbance scale of a spectrophotometer is the greatest accuracy obtained
when making a quantitative determination of a chromophoric substance?

_____

8. When you plot a graph of the absorbance (A) versus the concentration of a substance, how do you
know if Beer's law is obeyed?

_____

9. In what ionic form must iron exist to form a colored complex with the phenanthroline reagent?

_____

10. Even though the iron-phenanthroline complex forms in a solution with a wide pH range of 2 to 9, why is a lower pH of 3 to 4 more desirable?

_____

_____

Colorimetric Determination of Phosphorus and Iron. Wet Ashing

INTRODUCTION

During dry ashing, nonmetal elements are apt to be lost from the sample by volatilization (Expt. 3). Wet ashing avoids this difficulty, but is somewhat hazardous because of the necessity to use concentrated, strong acids and powerful oxidizing agents that can sometimes cause violent explosions. Ordinarily, the wet ashing process is applied to relatively small amounts of sample, usually only a few grams.

The qualitative detection, and particularly the accurate quantitative determination, of many of the biologically important mineral elements is a difficult matter because of the very small amounts involved. The concentrations of such elements as iron, copper, manganese, zinc, molybdenum, chromium, tin, cobalt, iodine, and selenium in biological materials typically range from a few hundred parts per million (ppm) to less than one part per billion (ppb). For this reason they are frequently called "trace elements." The analyses are commonly done by colorimetric methods, although more sophisticated procedures utilizing complex, expensive instruments are often more convenient. The latter techniques include atomic absorption spectrometry, atomic fluorescence spectrometry, emmission spectrometry, X-ray fluorescence spectrometry, and activation analysis. (See Ref. Pinta, 1978)

Phosphorus is an element required for the reproduction, growth, and metabolism of all living things. We include it in our fertilizers so it can replenish the soil, which supports plant life. Humans and animals need it for their bones and teeth. Phosphorus is present in the structure of many important biochemical compounds, frequently as part of an ester or anhydride group. For example, it forms a diester linkage with ribose or deoxyribose in the "backbone" of the nucleic acids, RNA and DNA. It is a part of the structure of sugar phosphates and acid anhydrides in metabolic pathways, and of phospholipids in many tissues. Thus when a biological material or food substance undergoes the wet ashing process, an appreciable amount is found in the ash solution in the form of inorganic phosphate. Phosphorus can then be determined colorimetrically by the Fiske and SubbaRow method. In this procedure, inorganic phosphate reacts with molybdic acid to form phosphomolybdic acid. The hexavalent molybdenum of this phosphomolybdate complex (but *not* the Mo of the excess molybdic acid) is reduced by a mixture of bisulfite and Elon (p-methylaminophenol) to form the "molybdenum blue" complex. The color intensity of the "molybdenum blue" is proportional to the original phosphorus content of the sample. The student should review the basic principles of colorimetry outlined in Experiment 4.

The determination of iron by the phenanthroline method is discussed in the previous experiment. Its estimation in the ash solution from biological materials, including foodstuffs, presents the possibility of interference from certain other ions. Chromium, zinc, and phosphate interfere if they exceed ten times the amount of iron; cobalt, copper, and nickel interfere more readily. A more detailed discussion of interfering ions and how to eliminate them are given in the references.

PROCEDURE

A. *Preparation of Sample by Wet Ashing.* (CAUTION. WEAR GOGGLES. Conc. sulfuric acid and 30% hydrogen peroxide cause skin burns. Handle carefully! Wash off spillage immediately with water.) The instructor will teach you the proper use of the analytical balance before you weigh your sample. See pages xiii-xiv.

Weigh a 1.000-g sample of finely ground, well-mixed material, place in a 125-ml Erlenmeyer flask and add 15 ml of concentrated H_2SO_4. Rotate the flask to mix the sample with the acid. Place a small, short-stem funnel in the neck, and heat the flask on a hot plate (or sand bath heated with a burner) for about 5 minutes after fuming starts. The temperature should be adjusted so that fumes are given off, but are not driven from the flask. Remove the flask from the hot plate, allow to cool for a few minutes, and add 5 ml of 30% hydrogen peroxide slowly, dropwise, to the sides of the funnel and flask. (**DANGER:** 30% H_2O_2 can cause explosions. Use a little at a time, and keep at arm's length, preferably behind a protective screen.) Slow addition in this manner avoids spattering and washes down charred material adhering to flask or funnel. Reheat the flask for about 2 minutes. If the material is still dark, cool the flask and add another ml of hydrogen peroxide as before. Reheat the flask.

This process of adding a ml of hydrogen peroxide and reheating is repeated until the solution is colorless, then the solution is heated slowly for 5 minutes to expel excess hydrogen peroxide. Removing the funnel prior to the last heating will aid in the expulsion of the hydrogen peroxide, but care in heating must be exercised to avoid loss of the liquid. Cool the flask to room temperature. Slowly and carefully add 20 ml of "extracting solution" (sodium acetate/acetic acid) with gentle swirling.

Filter the solution through an ashless filter paper (Whatman No. 40), using a glass funnel supported so that the tip of the funnel is just below the neck of a 50-ml volumetric flask. Rinse the 125-ml Erlenmeyer flask with 5 ml "extracting solution" and, *after* the filtering of the original solution is complete, add this rinse to the filter paper. Now rinse the filter paper with successive small volumes (2-3 ml) of "extracting solution" until the level of the solution in the volumetric flask has reached the lower end of the neck of the flask. Make to volume when flask is at room temperature and mix thoroughly. Label this "solution A."

B. *Determination of Phosphorus.* The standard curve to be used will cover the range of about 10 to 80 μg per tube. The aliquot of solution A (prepared in part A) to be pipetted for colorimetric analysis will depend on the phosphorus content of the original sample. If the phosphorus content was 100 mg P per 100-g sample, then wet ashing of a 1.0-g sample would result in having 20 μg P per 1 ml of solution A. This information can serve as a guide for judging the amount of solution A to use. If the approximate P content of the sample is not known, it is best to start with both 0.5-ml and 1.0-ml aliquots, and then repeat the analysis with some other volume if necessary.

Thoroughly clean, rinse, and dry eight 150-mm test tubes. (**NOTE:** Since many cleaning agents contain phosphates or polyphosphates, be sure that all glassware has been adequately rinsed.) Label the test tubes as indicated in the table and proceed as follows.

1. Pipet 0.50-ml and 1.00-ml aliquots from solution A using the 1-ml graduated pipet. *Do NOT pipet by mouth;* solution A is extremely acid (about 10 N H_2SO_4). Your instructor will show you the proper pipetting technique.
2. Pipet the volumes of standard P solution shown in the table, using a clean 1-ml graduated pipet. 1 ml = 100 μg P.
3. Make all tubes to 3.00 ml by adding the distilled water indicated, using the 10-ml graduated pipet.

	Blank	Standard P Tubes					Sample	
Tube No.	B	1	2	3	4	5	S-1	S-2
Ml solution A	---	---	---	---	---	---	0.50	1.00
Ml standard P	---	0.10	0.20	0.40	0.60	0.80	---	---
Ml distilled H_2O	3.00	2.90	2.80	2.60	2.40	2.20	2.50	2.00

4. Add exactly 1.00 ml of acid molybdate reagent to each tube from the 10-ml buret on the reagent bench. Mix.

5. Add exactly 5.00 ml distilled water to each tube from a 50-ml buret.

6. Add exactly 1.00 ml of 1% Elon/3% sodium bisulfite reducing reagent to each tube from the 10-ml buret provided. Mix thoroughly by complete inversions of test tubes (use Parafilm or rubber stoppers). **NOTE:** The color that develops is not stable. You should not add the 1% Elon unless you will be able to read the color intensities in 15 minutes. Check if an instrument will be available.

7. Let tubes stand for 15 minutes and then read the absorbance and %T against the reagent blank at a wavelength of 60 nm. Follow the "Notes on the Use of the Spectronic 20" on page 21.

8. If the absorbance reading of a sample is above 0.90, set up another sample. Do *NOT* dilute the colored solution with water, since serious errors may result. Conversely, if the absorbance reading of a sample tube is below 0.1, prepare a new sample using a larger aliquot of solution A. Enter all readings on the Laboratory Report, and calculate mg P per 100-g sample.

C. *Determination of Iron.* Because of the low iron content of foods and biological materials compared to phosphorus, a larger volume of sample solution A from part A will be used, and the iron standards will cover the range of 5 to 40 μg Fe per tube. If a sample contains 100 μg Fe per 1 g of original, weighed sample, this represents 2 μg Fe per 1 ml of prepared solution A. It is recommended, therefore, that 3- to 5-ml aliquots of solution A be tried initially.

Clean and dry 8 test tubes and label them: B (for blank), numbers 1 through 5, and S-1 and S-2 (for samples). Set up the tubes according to the steps given below. With the exception of the standard Fe solution, all reagents are available in 25-ml burets.

1. Pipet 3.0 and 5.0 ml of sample solution A into tubes S-1 and S-2, respectively. *Do not pipet by mouth suction.*

2. To tubes 1 through 5, respectively, add exactly 0.50, 1.00, 2.00, 3.00, and 4.00 ml of the standard iron solution (1 ml = 10 micrograms Fe). Use a 5- or 10-ml graduated pipet.

3. Add exactly 1.00 ml of hydroxylamine hydrochloride ($NH_2OH \cdot HCl$) solution to each tube including the blank. Mix.

4. Add exactly 1.00 ml of 1.5 M sodium acetate solution to each tube.

5. Add exactly 2.00 ml of 0.1% phenanthroline solution to each tube. Mix.

6. Add distilled water to each tube to give a final volume of 10 ml as follows:

Tube	B	1	2	3	4	5	S-1	S-2
ml H$_2$0	6.0	5.5	5.0	4.0	3.0	2.0	3.0	1.0

7. Cover each tube with a piece of Parafilm or a *clean* rubber stopper. Mix the contents of each by 5 or 6 complete inversions to ensure homogeneous solutions. Complete color formation occurs within 5-10 minutes.

8. Read the absorbance and the %T against the reagent blank at a wavelength of 508 nm. Record all readings on the report sheet and calculate mg Fe per 100-g sample.

REFERENCES

FRANSON, M. A., ed. 1976. *Standard methods for examination of water and wastewater.* Washington, D.C.: Amer. Public Health Assoc. Publications.

KOLTHOFF, I. M., and ELVING, P. J., eds. 1978. *Treatise on analytical chemistry, part I: theory and practice,* 2d ed., vol. 1. New York: John Wiley and Sons.

PINTA, M. 1966. *Detection and determination of trace elements.* Translated by M. Bivas. Ann Arbor, Mich.: Ann Arbor Science Publishers.

———. 1978. *Modern methods for trace element analysis.* Translated by STS, Inc. Ann Arbor, Mich.: Ann Arbor Science Publishers.

UNDERWOOD, E. J. 1971. *Trace elements in human and animal nutrition.* 3d ed. New York: Academic Press.

Reagents, Materials, and Equipment are listed on page 272.

LABORATORY REPORT

Name _____ Section _____ Date _____

Experiment 5. COLORIMETRIC DETERMINATION OF PHOSPHORUS AND IRON. WET ASHING

Record the absorbance and % transmittance for each tube in the tables below. Calculate the micrograms of P or Fe in each standard tube listed in the respective table. The amount of P or of Fe per tube will be obtained from the respective standard curves you will construct from the data.

Determination of Phosphorus

	Standard Tubes (100 μg P/ml)					Sample Solution A	
Tube No.	1	2	3	4	5	S-1	S-2
Ml P std./tube	0.1	0.2	0.4	0.6	0.8	––	––
μg P/tube							
Absorbance							
% Transmittance							

Determination of Iron

	Standard Tubes (10 μg Fe/ml)					Sample Solution A	
Tube No.	1	2	3	4	5	S-1	S-2
Ml Fe std./tube	0.5	1.0	2.0	3.0	4.0	––	––
μg Fe/tube							
Absorbance							
% Transmittance							

Standard Curve Prepare a standard curve by plotting absorbance on the ordinate (vertical axis) and the micrograms of P per tube on the abcissa (horizontal axis). Using the longer edge of the graph paper, try a scale where 10 μg P = 1 inch. On a separate graph paper, prepare a standard curve for iron, letting 10 μg Fe = 2 inches.

Calculations   Name of substance analyzed _____

	Phosphorus		Iron	
	S-1	S-2	S-1	S-2
μg/tube (from std. curve)	_____	_____	_____	_____
Ml solution A/tube	_____	_____	_____	_____
μg/1 g sample ashed	_____	_____	_____	_____
Mg/100 g sample	_____	_____	_____	_____

Name_____ Section_____ Date_____

Experiment 5. COLORIMETRIC DETERMINATION OF PHOSPHORUS AND IRON. WET ASHING

1. Compared to dry ashing, what are the principal advantages and disadvantages of wet ashing in preparing a sample for analysis?

   Advantages:_____

   _____

   Disadvantages:_____

   _____

2. a. List the chemical *symbols* of some of the physiologically important mineral elements.

   _____

   b. What makes the analysis of these elements so difficult?

   _____

3. List the important biochemical compounds that contain phosphorus.

   _____

   _____

4. a. How many grams of sodium acetate and how many ml of glacial acetic acid are present in 50 ml of "extraction solution" you used after wet ashing? (See Reagents in appendix.)

   _____

   b. Give an acceptable reason for the inclusion of so much sodium acetate.

   _____

5. Briefly outline the steps involved in the conversion of phosphate ions to "molybdenum blue."

   _____

   _____

   _____

6. Compare the *stability* of the colored complexes formed in the determination of phosphate and of iron.

   _____

   _____

NOTE: If your laboratory schedule did not include the performance of the laboratory procedures in Expt. 4, attach the answers to questions 7 through 10 from Expt. 4 to the question sheet of Expt. 5.

Acidity

Active Acidity. Measurement of pH

INTRODUCTION

Active acidity refers only to the hydrogen ions present in a solution due to the ionization of an acid. The percent of ionization of a strong, monoprotic acid approaches 100% (e.g., 0.1 M HCl is 92% ionized), while that of a weak acid is very low (e.g., 0.1 M acetic acid is 1.3% ionized). The hydrogen ion concentration of a monoprotic acid can be approximated from the molarity of the acid and its percent ionization, viz., $[H^+] = M \times \%/100$. Because of the stepwise ionization of diprotic and triprotic acids, the calculation is more complicated. In addition, the % ionization of an acid varies with its molarity, being greater in more dilute solutions. Further, the presence of the salt (conjugate base) of a weak acid will affect the extent of its ionization. Most biochemically important acids are *weak* acids.

Based on the ion product constant of water, $K_w = [H^+] \times [OH^-] = 1 \times 10^{-14}$ at 25°C, it is possible to calculate any one of the three values: $[H^+]$, $[OH^-]$, or pH, as long as one of these values is known.

Active acidity is commonly expressed in terms of the pH of a solution, where the pH represents the negative logarithm of the hydrogen ion concentration, viz., $pH = -\log [H^+]$. If the $[H^+]$ of any solution is: $a \times 10^{-b}$ moles per liter, then the pH can be calculated from the general equation: $pH = b - \log a$.

The pH of a solution can be determined by two general methods, (1) by means of pH indicator papers, and (2) by use of a pH meter. The pH papers are impregnated with organic dyes whose color is dependent on pH. Wide-range pH papers permit the estimation of pH within 1 pH unit, and narrow-range papers bring the pH estimation within ±0.5 unit. The pH meter accurately measures the pH of a solution within ±0.02 pH unit, and models with the expanded scale can determine the pH within ±0.002 unit under proper conditions. The calibrated pH scale has a pH range of 0 to 14, with subdivisions of 0.1 pH unit. The instrument consists of an electronic circuit, which measures the potential difference between a glass electrode and a calomel reference electrode.

Th glass electrode has a thin glass bulb at one end of a tube made of special types of glass. The bulb end of the tube contains a 0.1 M HCl solution. An inner silver-silver chloride electrode, which dips into the 0.1 M HCl solution, is sealed into the upper end of the glass tube. When the bulb of a glass electrode is immersed in a sample solution, the difference in the $[H^+]$ of the 0.1 HCl inside the bulb and the $[H^+]$ of the sample outside the bulb will result in the development of an electrical potential across the glass membrane. The pH of the sample solution is related to this potential difference. If a reference electrode is also immersed in the sample solution to complete the electrical circuit to the pH meter, this potential difference can be measured in terms of pH units. The calomel reference electrode consists of an inner electrode of mercury and calomel (mercurous chloride, Hg_2Cl_2) that dips into a saturated potassium chloride solution. Generally, a capillary or a wick in the bottom end of the tube permits the saturated KCl solution to make a "liquid junction" between the sample solution and the calomel inner electrode.

The availability of the glass electrode in the early 1940s as a simple and convenient method of measuring the pH of all types of aqueous solutions, including colored and turbid, was one of the most important advances in chemistry and the biological sciences. However, the fragility of the bulb at the end of the glass electrodes necessitates careful handling to avoid breakage during the lowering of the electrodes into a container or in the use of magnetic stirrers. Temperature changes affect the pH reading, so that some pH meters have a temperature control on the instrument panel, while others employ a correction graph, if the sample temperature is lower or higher than room temperature. Most glass electrodes will give accurate readings in the pH range of 0 to 9 or 10, but electrodes made of a special type of glass are required for alkaline solutions with high pH values.

All pH meters must be standardized with solutions of known pH values. For greater accuracy, the pH of the standard should be as close to the sample pH as is practical. However, a pH meter must be checked with at least two different standards (e.g., pH 4 and pH 7 or 8) when first put into operation, and occasionally thereafter during the day. Your instructor will demonstrate the use and the care of a pH meter.

Figure 6.1 illustrates a simple, reliable, easy-to-use pH meter. It has a pH scale only (no millivolt scale) and is calibrated to operate at room temperature (25°C). Figure 6.2 shows a research type of pH meter that has a millivolt scale permitting use of many other types of electrodes for oxidation-reduction potentials. It has a temperature compensator and can be attached to a recorder. It is illustrated with a single probe containing both a glass electrode and the calomel reference electrode that makes it possible to measure the pH of a solution in a test tube or other restricted areas where two separate electrodes would not fit. Still other new types of pH meters are becoming available with a digital readout of the pH value.

PROCEDURE

A. *Determination of pH with Hydrion Paper.* The instructor will demonstrate the use of the Hydrion paper as a group experiment with the students assisting in the estimation of the pH of the sample being tested. From the list of substances given below, four of them should be selected that cover the entire range, such as 0.1 M HCl, 0.1 M NaOH, and two other substances of intermediate pH. (Do the observed pH values for 0.1 M HCl and 0.1 M NaOH come close to their theoretical pH values if 100% ionization is assumed?) Later the student will have an opportunity to do several pH measurements with Hydrion paper in part B.

In making measurements with Hydrion paper, wide-range paper (pH 1-14) should be tried first in order to get an approximate idea of the pH. Subsequently, a suitable narrow-range paper should be used in order to determine the pH to within ±0.5 unit. For each individual pH determination, obtain a very small piece of the Hydrion paper, not more that 10 mm long, and place a small droplet of the liquid to be tested on it. The color developed should then be compared immediately with the appropriate standard color series for that particular pH range. If the color produced by the liquid on the Hydrion paper is similar to the color on either end of the color chart, the liquid being tested may not necessarily have the pH indicated. It is important to try the next *lowest* (or next *highest*) test range to obtain the correct pH.

Record these pH results on the report sheet.

B. *Measurements with the pH Meter.* The instructor will demonstrate the proper use of the pH meter. Note that the ends of the reference and glass electrodes are immersed in a beaker of water when not in use, and the "Function" knob is in the "Standby" position. Be sure the calomel reference electrode is vented. Usually the instructor takes care of the uncovering and covering of the vent hole. When using the pH meter, the following steps should be observed.

1. Raise the electrode head on the mounting post until the two electrodes are above the top of the beaker of water. Rinse the lower half of each electrode with a stream of water from your wash bottle. Blot the tips of the electrodes with a tissue.
2. Place a 50- or 100-ml beaker, filled one-third full with pH 4.00 or pH 7.00 standardizing solution, under the electrodes. Use standardizing solution with pH closest to anticipated pH of sample solution.
3. *Carefully* lower the electrodes into the solution until the bulb of the glass electrode is under the surface of the solution. (**CAUTION:** Avoid striking the glass electrode on the edge of the beaker or you might crack or break this expensive electrode.)

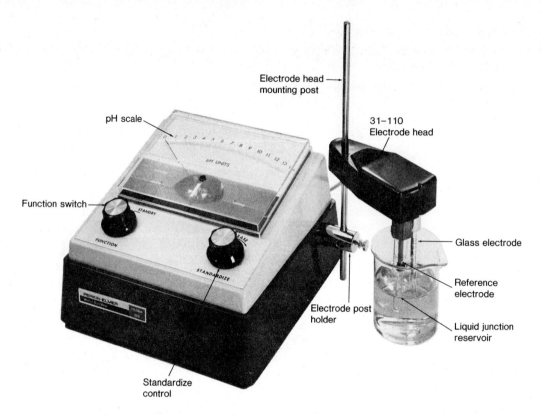

Figure 6.1. Coleman METRION III pH Meter. (Courtesy of Coleman Instrument Division of the Perkin-Elmer Corporation)

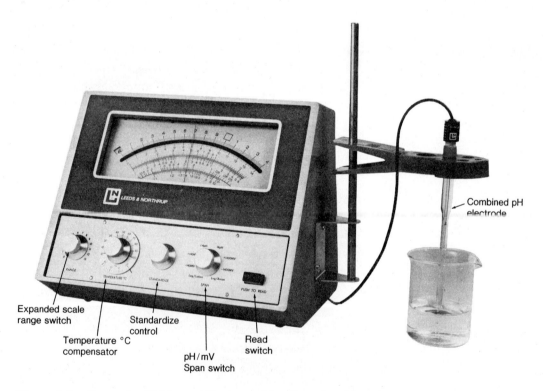

Figure 6.2. Research pH Meter. (Courtesy of Leeds and Northrup)

4. Turn the "Function" switch to "pH."

5. Adjust the "Standardize" control until the meter needle exactly reads the pH of the standardizing solution.

6. Turn the "Function" switch to "Standby."

7. Raise the electrodes from the standardizing solution, cover it with a watch glass and set aside.

8. Replace the beaker of water underneath the electrodes and rinse the two electrodes with a stream of water from wash bottle. Wipe electrodes with a tissue. *Do not rinse water into the standard solution.*

9. Lower the electrodes carefully into the sample solution of unknown pH.

10. Turn "Function" switch to "pH" and read the pH of the sample solution. Estimate the second decimal place.

11. Return switch to "Standby." Remove sample solution and *immerse the electrodes in the beaker of water.*

NOTE: If you are uncertain that the electrodes or the pH meter are functioning normally, standardize the instrument with *both* standard pH solutions. After setting the meter with one of the standard solutions, the pH meter should read within $\pm$ 0.1 pH unit of the pH of the second standard when the latter solution is read against the first setting without readjusting the standardization knob.

Unknown sample. You will be given 20-25 ml of an unknown solution in a test tube. *First* determine the approximate pH of the sample with *Hydrion* paper, starting with the wide-range paper and then using the narrow-range Hydrion paper. Finally standardize a pH meter with the standard pH solution (pH 4 or 7), which is closest to the approximate Hydrion pH of your unknown. Transfer your unknown solution to a small beaker and read the exact pH of your unknown. Record your pH values on the report sheet.

pH of common substances. From the list of common substances given below, select at least three and determine first the approximate pH with Hydrion paper, and then a more accurate value with the pH meter. Record your pH values. Do not repeat any substance used in part A. Insofar as possible, the liquids to be tested should be brought to the laboratory by the individual student. If a substance is highly colored, the use of Hydrion paper may not be successful.

fresh milk	blood	distilled water
sour milk	urine	distilled water, boiled
orange juice	saliva	0.1 M HCl
lemon juice	beer	0.1 M NaOH
grapefruit juice	wine	0.1 M Na_2HPO_4
tomato juice	city water	0.1 M Na_2CO_3
vinegar	lake water	0.1 M $NaHCO_3$

REFERENCES

BATES, ROGER G. 1973. *Determination of pH: theory and practice.* 2d ed. New York: John Wiley and Sons.

———. 1978. Concepts and determination of pH in *Treatise on analytical chemistry,* ed. I. M. Kolthoff and P. J. Elving, 2d ed., vol. 1, part I, chapter 14. New York: John Wiley and Sons.

Reagents, Materials, and Equipment are listed on page 273.

LABORATORY REPORT

Name_____ Section _____ Date _____

Experiment 6.   ACTIVE ACIDITY. MEASUREMENT OF pH

A. *Measurement of pH with Hydrion Paper*

Indicate the pH of at least 4 substances you tested with Hydrion paper.

Name of substance tested	pH
0.1 M HCl	_____
_____	_____
_____	_____
0.1 M NaOH	_____

B. *Measurements with the pH Meter*

	pH measurement using	
	Hydrion paper	*pH meter*
Unknown Solution No. _____	_____	_____
Common Substances		
_____	_____	_____
_____	_____	_____
_____	_____	_____

QUESTIONS AND PROBLEMS

Name_____ Section_____ Date_____

Experiment 6.   ACTIVE ACIDITY. MEASUREMENT OF pH

1. A piece of Hydrion paper is moistened with a sample and the color of the moist area matches the end block of the color chart. Is the pH designated by this color block to be accepted as the pH of the sample? Explain.

   _____

   _____

2. How closely can the pH of a solution be determined

   a. by Hydrion paper?_____   b. by a pH meter? _____

3. Briefly describe the structure and the chemical contents of the

   a. Glass electrode. _____

   _____

   b. Calomel reference electrode. _____

   _____

   _____

4. Assuming 100% ionization, calculate the pH of 0.1 M HCl and 0.1 M NaOH solutions. Compare your values with the estimated pH obtained when using Hydrion paper in part A. Show work.

   a. 0.1 M HCl   b. 0.1 M NaOH

   pH with Hydrion paper: 0.1 M HCl _____   0.1 M NaOH _____

5. Which one has the greater active acidity, 0.1 M HCl or 0.2 M acetic acid? Compare by calculating the $[H^+]$ of each solution using the % ionization given in the introduction to the experiment. Show work.

   a. 0.1 M HCl _____

   b. 0.2 M CH_3COOH_____

   c. Which is more acid and how many times more acid?

   _____

6. How many times as acid is the first compared to the second of each of the following?

   a. Gastric juice at pH 2 vs. intestinal juice at pH 8.   _____

   b. An orange at pH 3 vs. a banana at pH 6.   _____

   c. Fresh milk at pH 7 vs. sour milk at pH 5.   _____

Buffers. The Control of pH

INTRODUCTION

The maintenance of an optimum pH in the environment of plant and animal cells is vital to the life of an organism. The control of pH, essential for enzyme-catalyzed reactions, both *in vivo* and *in vitro*, is one of the main functions of buffers. The control of pH of blood close to pH 7.4 by its buffer systems is necessary for good health, and even for life itself, because the oxygen-carrying capacity of hemoglobin is markedly reduced if the pH drops more than a few tenths.

A buffer is a mixture of a weak acid and its conjugate base. The dissociation of a weak acid in an aqueous solution can be represented by the equilibrium equation: $HA \leftrightarrows H^+ + A^-$. While this oversimplified equation represents a monoprotic acid, there are times when HA is an ion (e.g., HCO_3^-, $H_2PO_4^-$, etc.) and occasionally the conjugate base is a molecule (e.g., NH_3, Tris). The following equations illustrate this point:

$$weak\ acid \leftrightarrows H^+ + conjugate\ base$$
$$acetic\ acid \quad CH_3COOH \leftrightarrows H^+ + CH_3COO^- \quad acetate\ ion$$
$$phosphoric\ acid \quad H_3PO_4 \leftrightarrows H^+ + H_2PO_4^- \quad dihydrogen\ phosphate\ ion$$
$$dihydrogen\ phosphate\ ion \quad H_2PO_4^- \leftrightarrows H^+ + HPO_4^= \quad monohydrogen\ phosphate\ ion$$
$$ammonium\ ion \quad NH_4^+ \leftrightarrows H^+ + NH_3 \quad ammonia\ (a\ molecule)$$

When the dissociation of the weak acid has reached an equilibrium point, the dissociation or ionization constant, K_a, can be written as follows:

$$K_a = \frac{[H^+] \times [A^-]}{[HA]}$$

With only the weak acid present in solution, $[H^+] = [A^-]$, and the $[H^+]$ can be obtained from the % ionization, if known, or from the pH of the solution. Because the weak acid usually has a low degree of dissociation, it is assumed that [HA] is equal to the initial molarity of the weak acid. Most K_a values of weak acids have been determined, and the magnitude of the K_a value is indicative of the strength of the acid. The larger the value of the constant, the greater is the ionization of the acid. For example, an acid with a $K_a = 1 \times 10^{-4}$ is more dissociated (stronger) than an acid with a $K_a = 1 \times 10^{-5}$.

A more convenient and useful way of expressing the degree of dissociation or strength of an acid is to use a term called the pK_a of a weak acid, which is the negative logarithm of the K_a, viz., $pK_a = -\log K_a$. In general terms, if the $K_a = c \times 10^{-d}$, then $pK_a = d - \log c$. For example, the dissociation constant for acetic acid is $K_a = 1.86 \times 10^{-5}$. Therefore, the pK_a value is calculated from the general equation as follows: $pK_a = 5 - \log 1.86 = 5 - 0.27$, or $pK_a = 4.73$. The *larger* the pK_a value, the *weaker* the acid. For example an acid with a $K_a = 1 \times 10^{-4}$ will have a $pK_a = 4$; a weaker acid with a $K_a = 1 \times 10^{-5}$ has a $pK_a = 5$.

The pK_a values of weak acids are necessary for calculating the pH or the ratio of conjugate base to acid in buffer solutions. These relationships are expressed in the Henderson-Hasselbalch equation, which can be developed from the dissociation constant, K_a, of a weak acid as follows:

$$K_a = \frac{[H^+] \times [A^-]}{[HA]}$$

rearranging, we get

$$[H^+] = K_a \times \frac{[HA]}{[A^-]}$$

and taking the negative logarithm ($-\log$) of both sides gives

$$-\log [H^+] = -\log K_a -\log \frac{[HA]}{[A^-]}$$

By substituting pH and pK$_a$ in the appropriate places and by inverting the ratio of the *last term* to give a $+$log ratio, we now have the usual form of the *Henderson-Hasselbalch equation:*

$$pH = pK_a + \log \frac{[A^-]}{[HA]} \quad \text{or,} \quad pH = pK_a + \log \frac{[conj.\ base]}{[weak\ acid]}$$

Many texts use the term *salt* for *conj. base*. While the conjugate base frequently is a salt, this is not always true, such as in the case of NH_4^+/NH_3 and $TrisH^+/Tris$ buffers, where the conjugate bases are molecules and *not* salts and the weak acids are ions of their corresponding salts (see second paragraph on page 49).

An inspection of the Henderson-Hasselbalch equation shows that when the molar concentrations are equal, the logarithm of their ratio is zero ($\log 1 = 0$), and the pH of the buffer solution equals the pK$_a$ value of the acid component. The useful pH range of a buffer in opposing a change in pH of a solution is about equal to the pK$_a \pm 1$. When the ratio of conjugate base to weak acid is 10/1, then $\log 10 = 1$, and the pH of the solution becomes equal to the pK$_a + 1$. Conversely, if the ratio is 1/10, the pH = pK$_a - 1$.

In general, when choosing a buffer mixture which will control the pH of a solution, a weak acid is selected whose pK$_a$ is as close as possible to the desired pH.

In controlling the pH of an aqueous medium, the buffer components must resist a change in pH when additional H^+ or OH^- are added to or produced in the medium. Each component functions as follows:

(1) weak acid: $HA + OH^- \rightarrow H_2O + A^-$

(2) conjugate base: $A^- + H^+ \rightarrow HA$

In (1), some of the weak acid is consumed and more conjugate base is generated; in (2), some of the conjugate base is used and more weak acid is formed. In either case, the pH of the buffered solution remains relatively constant when the added H^+ or OH^- react with the buffer components. However, as a result of the reactions that occur in (1) or (2), the ratio of conjugate base to weak acid will be changed somewhat and will slightly change the pH of the solution in accordance with the Henderson-Hasselbalch equation. The extent of the pH change depends on the buffering capacity of the solution.

The ability of a buffer system to resist a change in pH is called its *buffer capacity*. This effectiveness of a buffer to do its work depends mainly on (1) the total concentration of the buffer components, and (2) the ratio of the conjugate base to the weak acid. It should be readily apparent that a 0.1 M buffer will have twice the buffering capacity compared to a 0.05 M buffer. The effect of the ratio of the conjugate base to the acidic component, however, is more variable. When the ratio of these buffer components is one (50% of each), the buffered solution can resist a pH change equally as well with the addition of either H^+ or OH^- within the limits of molar capacity. When the concentration of the *conjugate base is greater* than that of the weak acid, the buffer system is more effective toward the addition of H^+ than OH^-. When the concentration of the *weak acid is greater* than that of its conjugate base, then the reverse is true. When more H^+ or OH^- are added to the buffered solution than can be counteracted by the appropriate component of the buffer system, the pH will change dramatically. You will have the opportunity to observe the effectiveness of your buffer in resisting such pH changes in part B of the experiment.

PROCEDURE

A. *Preparation of a Buffer.* You will be assigned one of the buffers listed in the table below. From the desired pH value of your buffer and the pK_a value of its acidic component, calculate the ratio of conjugate base to acid, using the Henderson-Hasselbalch equation. From this ratio, determine the volumes of 0.1 M solutions of conjugate base and acid to prepare a *total* volume of 90-100 ml of buffer solution.

In calculating the ratio of the conjugate base to acid from the Henderson-Hasselbalch equation in the form as derived above, the log of the ratio will be a *negative value* if the desired buffer pH is *less than* the pK_a value of the acid. This negative value makes it more difficult for most students to determine the corresponding "antilog," which is the actual ratio of conjugate base to weak acid. To avoid this situation, the ratio is inverted and the sign in front of the log is changed: viz., $+\log b/a = -\log a/b$. Therefore, depending on the relative values of your desired pH and the pK_a of the acid, select one of the following two forms of the Henderson-Hasselbalch equations for your calculations. Do not forget where the "conjugate base" and "weak acid" terms are located in your ratio.

Desired pH is greater than pK_a

$$pH = pK_a + \log \frac{[\text{conj base}]}{[\text{weak acid}]}$$

Desired pH is less than pK_a

$$pH = pK_a - \log \frac{[\text{weak acid}]}{[\text{conj base}]}$$

In this experiment you can express the ratio of the conjugate base and weak acid as *whole* numbers. Supposing the [conj. base]/[weak acid] is 2 (the antilog of 0.30). This actually means that the ratio is 2 to 1, or 2/1. The concentration units generally are expressed in "moles," but because you are preparing a small volume of buffer, you will use "millimoles" (abbrev. mmoles). The number of mmoles $= $ ml $\times$ M, and since the molarities of all the reagents used to prepare the buffers listed in the table below are 0.1 M, the ratio of conjugate base to weak acid that you determine can be simply expressed in milliliters, as shown below, using the 2/1 ratio as an example:

$$\frac{[\text{conj base}]}{[\text{weak acid}]} = \frac{2 \text{ mmoles}}{1 \text{ mmole}} = \frac{2(\text{ml}_b \times \cancel{0.1\text{ M}})}{1(\text{ml}_a \times \cancel{0.1\text{ M}})} = \frac{2 \text{ ml}_b}{1 \text{ ml}_a}$$

In order to prepare 90-100 ml of the buffer solution, it is necessary to multiply both the numerator and denominator of your volume ratio by some number so that the sum of the two volumes falls in the 90-100-ml range.

Check your calculations with your instructor *before* you actually prepare your buffer. All of the 0.1 M reagents listed in the table below are available on the reagent shelf. *Read the labels carefully* to avoid mistakes and waste. Measure the required volumes of 0.1 M reagents with your 100-ml graduate and mix in a 250-ml beaker. Measure the pH of your buffer with a standardized pH meter as you were previously instructed in Experiment 6, taking care not to damage the bulb of the glass electrode. Bring your buffer to the instructor for a final check on the instructor's pH meter before proceeding to part B. Record all data on your report sheet.

B. *Buffer Capacity of Your Buffer.* Review the discussion on buffer capacity in the introduction.

Pipet 10.0 ml of your buffer into each of eight clean test tubes. Into four of these test tubes pipet 1, 3, 6, and 9 ml of 0.1 M HCl, respectively. Into the remaining four test tubes pipet 1, 3, 6, and 9 ml of 0.1 M NaOH, respectively. (Do *not* measure these volumes with a graduated cylinder.) Mix thoroughly. Measure the pH of each of the four tubes containing HCl with a standardized pH meter by pouring one tube at a time into a 50-ml beaker. You can make the successive readings by draining and refilling the beaker *without rinsing*, provided you test the four tubes in increasing order of HCl additions. (**NOTE:** The bulb of the glass electrode does not have to be completely immersed in the solution. As long as the tip of the bulb makes contact, you will get a correct reading. You may have to tilt the beaker slightly with the first solution to make contact. In any event, be careful in using the glass electrode in small beakers. If the combination, single electrode illustrated in fig. 6.2, Expt. 6, is available, the pH readings can be made directly in the test tube.)

Thoroughly rinse the 50-ml beaker and take the pH readings of the four tubes containing the added 0.1 M NaOH in the same manner as above.

Record the pH readings in the table on your report sheet and calculate the millimoles of the substances requested.

C. **Buffer Capacity of Distilled Water.** Your instructor will provide the class with a bottle of previously boiled and cooled distilled water.

Pipet 10.0 ml of this water into two separate test tubes. To one tube add 1.0 ml of 0.1 M HCl and to the other tube add 1.0 ml of 0.1 M NaOH. Mix each thoroughly. Measure the pH of the boiled and cooled distilled water and the two tubes containing the added HCl and NaOH with the pH meter. Record the results and compare the pH changes with those when 10-ml buffer was treated with 1 ml of 0.1 M HCl and NaOH.

Data for the Preparation of Buffers

Buffer number	Desired pH	Name of buffer system, acid, and conjugate base	pK_a of acidic component
1 2	2.40 2.58	primary phosphate H_3PO_4, $H_2PO_4^-$	2.10
3 4 5 6 7	4.22 4.40 5.00 5.18 5.30	acetate CH_3COOH, CH_3COO^-	4.70
8 9 10 11 12	6.20 6.32 6.50 7.10 7.28	secondary phosphate $H_2PO_4^-$, HPO_4^{-2}	6.80
13 14 15 16 17	7.62 7.80 8.40 8.58 8.70	tris(hydroxymethyl)aminomethane Tris H^+, Tris $(HOCH_2)_3CNH_3^+$, $(HOCH_2)_3CNH_2$	8.10
18 19 20	8.82 9.00 9.60	ammonia NH_4^+, NH_3	9.30

REFERENCE

Suttie, J. W. 1977. *Introduction to biochemistry.* 2d ed. New York: Holt, Rinehart & Winston, chapter 1.

Reagents, Materials, and Equipment are listed on page 274.

LABORATORY REPORT

Name_____Section_____ Date _____

Experiment 7. BUFFERS. THE CONTROL OF pH

A. *Preparation of a Buffer*

Buffer No. _____

Name and formula of buffer system_____

Desired pH _____   pK$_a$ of acidic component _____

Calculate the ratio of conjugate base to acid. Show specific chemical formulas used and appropriate substitutions into the Henderson-Hasselbalch equation.

	Reagents used	*Volume*	*Molarity*
Conjugate base:	_____	_____	_____
Weak acid:	_____	_____	_____

pH of your buffer on the pH meter. _____

B. *Buffer Capacity of Your Buffer*

Millimoles buffer components in 10 ml of 0.1 M buffer solution. _____

Ratio of conjugate base to weak acid. _____

(a) Millimoles conj. base/10 ml = _____   (b) Millimoles weak acid/10 ml = _____

	Addition of 0.1 M HCl			Millimoles after HCl addition		Calculated pH of buffer*
ml added	(c) mmoles	pH of buffer		(a) − (c) conj. base	(b) + (c) weak acid	
1						
3						
6						
9						-----

*Use Henderson-Hasselbalch equation

	Addition of 0.1 M NaOH			Millimoles after NaOH addition		Calculated pH of buffer*
ml added	(d) mmoles	pH of buffer		(a) + (d) conj. base	(b) − (d) weak acid	
1						
3						
6						
9						–––––

*Use Henderson-Hasselbalch equation

With some of the prepared buffers the pH changes readily upon the addition of 6 ml of 0.1 M HCl or 0.1 M NaOH. Why does the buffer exhibit such poor "buffering ability" in such instances?

Write balanced *net ionic* equations showing how your buffer acts to oppose a change in pH

1. Upon the addition of acid (H^+):_____

2. Upon the addition of base (OH^-):_____

C. **Buffer Capacity of Distilled Water.**

What was the pH of boiled and cooled distilled water?_____
Compare the buffer capacity of distilled water versus your buffer after adding 1 ml of 0.1 M HCl or 0.1 M NaOH.

	0.1 M HCl	0.1 M NaOH
pH of distilled water	_____	_____
pH of your buffer (part B above)	_____	_____

Complete the following table: (assume 100% ionization of HCl and NaOH)

	Observed pH	Calculated molarity	[H^+]	Calculated pH
10 ml distilled water + 1 ml 0.1 M HCl	_____	_____	_____	_____
10 ml distilled water + 1 ml 0.1 M NaOH	_____	_____	_____	_____

What did your observations tell you about the buffering capacity of water?

QUESTIONS AND PROBLEMS

Name_____ Section_____ Date _____

Experiment 7. BUFFERS. THE CONTROL OF pH

NOTE: Show the mathematical steps required to solve each problem and the units of volume, weight, etc., associated with the numerical values in the data and answers.

1. Write the formula of the conjugate base for each of the following:

   Carbonic acid _____   Glycine ($CH_2NH_3^+COO^-$)_____

   Formic acid_____   Bicarbonate ion _____

2. Calculate the pK_a value of an acid whose K_a value is given below.

   (a) $K_a = 4 \times 10^{-5}$
      Ans. $pK_a = 4.4$

   (b) $K_a = 6.3 \times 10^{-8}$
      Ans. $pK_a = 7.2$

3. What is the K_a and pK_a of lactic acid if a 0.05 M solution has $[H^+] = 2.6 \times 10^{-3}$? Write the ionization equation for lactic acid.
   Ans. $K_a = 1.35 \times 10^{-4}$, $pK_a = 3.87$

4. Under what conditions will the pH of a buffer solution equal the pK_a value of the acidic component?

   _____

5. In opposing a pH change, which buffer component reacts

   With H^+ ions?_____   With OH^- ions?_____

6. When the buffer components react with added H^+ or OH^- ions, does the ratio of conjugate base to weak acid, and the pH of the buffer, remain unchanged? Explain.

   _____

   _____

7. a. How does dissolved CO_2 in distilled water lower the pH of *unboiled* water? Illustrate with equations.

   _____

   b. Why does boiling the water before using it in part C raise the pH of the water closer to neutral (pH 7)?

   _____

   _____

8. If 400 ml of 0.1 M NaH_2PO_4 is mixed with 100 ml of 0.1 M H_3PO_4, what is the pH of the buffer mixture? The pK_a of H_3PO_4 is 2.1. Show Henderson-Hasselbalch equation with chemical formulas of buffer components *before* substituting numerical values. Ans. pH = 2.7.

9. How would you prepare 100 ml of 0.1 M NH_4^+/NH_3 buffer of pH 9.78 from 0.1 M solutions of the reagents? The pK_a of NH_4^+ is 9.30. Show work as indicated above in (8).

Ans. 75 ml 0.1 M NH_3 (e.g., NH_4OH) and 25 ml 0.1 M NH_4Cl.

Total Acidity

INTRODUCTION

The total acidity of a solution refers to both the ionized and undissociated acid present in the solution. Total acidity is determined by the titration process in which a standard solution of a base ($NaOH$ or KOH) is added from a buret to the acidic solution contained in an Erlenmeyer flask or beaker in the presence of an indicator (an organic dye). When the indicator changes color, the *equivalence point,* or "end point," has been reached and the titration is complete. The titration of a basic substance in solution is carried out similarly with a standard solution of an acid (usually H_2SO_4 or HCl). In the neutralization reaction between an acid and a base, the actual ionic reaction involves the hydrogen ions and the hydroxide ions to form water, viz., $H^+ + OH^- \rightarrow H_2O$.

The concentration of a standard solution is expressed in terms of its *normality.* A 1 N solution of an acid contains 1 equivalent weight (1 mole) of H^+ per liter of solution, and a 1 N solution of a base contains 1 equivalent weight (1 mole) of OH^- or other proton-accepting substance per liter of solution. The units for normality are "equivalents per liter" or "milliequivalents per milliliter" (i.e., 1 N = 1 eq/liter or 1 meq/ml). The *gram-equivalent weight of an acid* is the amount of an acid in grams that furnishes 1 mole (1 g) of H^+ and can be calculated as follows:

$$\text{Gram-equivalent wt of acid} = \frac{\text{gram-formula weight of acid}}{\text{no. of ionizable } H^+ \text{ per formula}}$$

The *gram-equivalent weight of a base* is the amount of a base in grams that contains 1 mole of OH^- or other proton acceptor (e.g., NH_3, Tris, etc.) and can be calculated from the equation:

$$\text{Gram-equivalent wt of base} = \frac{\text{gram-formula weight of base}}{\text{no. of } OH^- \text{ per formula (or other } H^+ \text{ acceptor)}}$$

Normalities of acids and bases used in biochemistry are generally 0.1 or less. The number of equivalents of an acid or base in a solution can be obtained by multiplying the normality by the volume in liters, that is, $N \times l =$ equivalents. Smaller volumes and lower concentrations of solutes are encountered in biochemical analysis, so that the *milliequivalent* (meq) is the preferred unit. The number of milliequivalents of an acid or a base can be calculated from the equation, $ml \times N =$ meq, where the units of normality are defined above. At the equivalence point of a titration, the meq of acid exactly equals meq of base ($meq_a = meq_b$), and this can be expressed in the equation: $ml_a \times N_a = ml_b \times N_b$. This relationship will be more apparent to the student in part A of this experiment. Furthermore, if the normality of an acid or base is not known, it can be calculated from the above mathematical relationship, provided the normality of one solution is known. A more detailed discussion of normality, preparation of solutions of any normality, and the relationship between normality and molarity of solutions can be found in most general chemistry textbooks.

PROCEDURE

A. *Practice in Titration.* You will be furnished with a standard 0.100 N H_2SO_4 solution and a standard NaOH solution whose normality will be near 0.1 N. The exact normality of the NaOH will be given to you. Carefully read the instructions given below on the proper use and care of burets, and follow suggestions given by your instructor.

Set up two burets in a buret clamp attached to a ring stand. The 25-ml buret will be used for the 0.1 N H_2SO_4 and the 50-ml buret for the NaOH solution. If the buret has a glass stopcock, it must be properly lubricated with stopcock grease so that it will turn easily. Always exert a slight pressure *toward* the buret when turning a glass stopcock; never pull on it, because solution will leak from around the loose stopcock, causing serious errors. Proper manipulation of the glass stopcock with the *left* hand will prevent a leaking buret, a common cause of analytical errors by students. The newer, but more expensive, Teflon stopcocks are self-lubricating and do not require grease. If a plastic nut holds the Teflon stopcock in place, it must be properly tightened to prevent leaking.

A buret must always be rinsed several times with small portions (about 5 ml) of the solution to be used to remove the traces of water in a wet buret or to remove possible contaminants from an apparently clean, dry buret. After rinsing, the buret is filled with reagent well above the zero mark, and the stopcock is opened to fill the buret tip. Be sure no air bubbles remain in the tip. Carefully note the graduation markings on your buret. The smallest division is 0.1 ml, and the buret reading can be estimated to the closest 0.01 ml. Most titration errors can be traced to the improper reading of a buret, titrating past the end point, leaky burets, and dirty burets. Small droplets of reagent adhering to the inside wall of a buret indicate improper drainage and, if excessive, can result in incorrect buret readings. If cleaning of a buret is necessary, consult your instructor.

After careful reading of the above instructions and observing the demonstration given by your instructor, fill your 25-ml buret with 0.100 N H_2SO_4 and the 50-ml buret with the standard NaOH solution. (Be sure stopcocks are closed.) Open stopcock to fill buret tip and to remove all air bubbles. Adjust the height of the solution in the buret so that the liquid meniscus is on the zero mark. The reading is facilitated by looking through the buret toward the light and placing your finger against the back of the buret about 1 cm *below* the meniscus. Proceed with the three trials given below.

Trial 1. Add a definite volume (14-16 ml) of 0.1 N H_2SO_4 from the 25-ml buret to a 125-ml Erlenmeyer flask. Record the *exact* volume in your lab manual. Add 3 drops 1% phenolphthalein and rinse inside wall of flask with distilled water (wash bottle). Titrate the acid solution with standard NaOH from the 50-ml buret until a *faint* pink color persists for 30-60 seconds, mixing well during the titration. As you approach the end point, the localized pink color will not disappear as rapidly upon swirling the flask. It is then advisable to add the NaOH solution dropwise, or even in fractions of a drop by touching the tip of the buret, containing a partial drop, to the inside surface of the neck of the flask. It is also good technique to wash down the neck of the flask with a stream of water just before the end point is reached.

An indicator blank correction is not necessary when using phenolphthalein indicator with 0.1 N solutions of strong acids and bases. However, this may not be true when using other indicators or other solutions, and an indicator blank correction may be required.

Record the ml of acid and base to the closest 0.01 ml on your laboratory report sheet and calculate the milliequivalents (meq) of each reagent using the formula: meq = ml $\times$ N. Express your results to *three decimal places*.

Trial 2. Repeat the above, but use 18-20 ml of 0.1 N H_2SO_4.

Trial 3. Repeat the above with 15-20 ml of 0.1 N H_2SO_4, but deliberately *overtitrate* by adding 1-2 ml NaOH in excess so that the indicator will be definitely red. Now back-titrate with 0.1 N H_2SO_4 to decolorize the solution, and then add standard NaOH to give a faint pink color. Record the *total* volumes of acid and base used and calculate the milliequivalents (meq) of each. This procedure illustrates how an overtitrated sample can be saved without discarding it.

In each of the above titration trials, the *difference* between meq H_2SO_4 and meq NaOH should not exceed 0.025 meq. If you use good technique, you should have no difficulty achieving results within this limit. *Your instructor must check your results for part A before you proceed further.*

B. *Total Acidity or Alkalinity of Various Substances.* The instructor will demonstrate the proper technique in the use of pipets *before* you proceed with this part of the experiment. If solid substances are included, then instructions on the use and care of the analytical balance are also required.

Each student will titrate household ammonia to become familiar with an indicator other than phenolphthalein, and to gain experience for a later determination of protein nitrogen by the Kjeldahl method (Expt. 27). In addition, one or two other materials will be assigned (see bulletin board). Note carefully the appropriate standard acid or base and indicator used with each assigned material as shown in the following table.

Material	Approximate amount used	0.1 N Reagent	Indicator	Component sought Name	Eq wt
Vinegar	5 g	NaOH	Phenolphthalein	Acetic acid	60
Lemon or lime juice	5 g	NaOH	Phenolphthalein	Citric acid	64
Grapefruit juice	10 g	NaOH	Phenolphthalein	Citric acid	64
White grape juice	10 g	NaOH	Phenolphthalein	KHtartrate	188
Cream of tartar	0.3 g	NaOH	Phenolphthalein	KHtartrate	188
Baking soda or an antacid (Tums)	0.2 g	H_2SO_4	Methyl orange	$NaHCO_3$	84
Household ammonia (diluted 1:10)	5 ml (0.5 g)	H_2SO_4	Methyl red – methylene blue	Ammonia	17

(1) *Liquid materials.* Assume that the density of each liquid material is 1.0 (1 ml = 1 g). Check the above table to determine whether you need 5 ml (5 g) or 10 ml (10 g) of the assigned material. The household ammonia has been diluted 1 to 10 for you (1 vol. NH_3 + 9 vol. H_2O), so that a 5-ml portion will contain 0.5 g of the *original* household ammonia.

Into two separate, rinsed 125- or 250-ml Erlenmeyer flasks, pipet exactly the required volume of sample, touching the tip of the pipet to the surface of the liquid or to the inside wall of the flask before withdrawing the pipet. Add distilled water to the flask to make a total volume of 40-50 ml. Add 3 drops of the appropriate indicator and titrate with standard NaOH (or H_2SO_4 for ammonia) in a manner previously described in part A. The duplicate titrations of the sample should agree within ±0.3 ml; if they do not, repeat a third time.

In the titration of household ammonia with H_2SO_4 in the presence of methyl red-methylene blue indicator, the color change at the end point is from green in alkali to a reddish-purple in acid with an intermediate slate-gray. To help you recognize the end point, set up two color-comparison flasks, each containing 3 drops of methyl red-methylene blue indicator in about 60 ml of distilled water. Make one flask alkaline (green) with a drop or two of dilute household ammonia and make the other flask acidic (reddish-purple) with 0.1 N H_2SO_4 added dropwise. This indicator is particularly useful for titrations involving weak bases and strong acids.

(2) *Solid materials.* Weigh two separate samples on an analytical balance on glassine paper to within 1 mg. Quantitatively transfer each sample to a labeled 125- or 250-ml Erlenmeyer flask and dissolve the material in 40-50 ml distilled water, warming gently if necessary. Add 3 drops of the appropriate indicator and titrate with NaOH or H_2SO_4 as specified in the above table.

If you are analyzing baking soda or antacid, the color change of the methyl orange indicator is rather subtle at its end point. To help you detect this color change during a titration, prepare two comparison flasks by adding about 60 ml of distilled water to each of two Erlenmeyer flasks. Add the same amount (3-4 drops) of indicator as used in the titration of the baking soda sample. Make one color flask distinctly yellow by adding one drop of 0.1 N NaOH; make the other color flask distinctly orange by adding 0.1 N H_2SO_4 dropwise (4-6 drops). Titrate the baking soda solution so that the end point color is intermediate (yellow-orange) to the colors of the two blanks. For more accurate results an indicator blank should be determined and subtracted from the sample titration.

(3) *Calculations.* Record the data obtained and complete the table on the laboratory report sheet. Record the values in columns 4 through 8 in three significant figures. The following relationships are used.

$$\text{meq of substance sought} = \text{meq of reagent used (col. 5)}$$
$$\text{wt of substance (col. 7)} = \text{meq of substance (col. 5)} \times \text{meq wt of substance (col. 6)}$$
$$\% \text{ found (col. 8)} = \frac{\text{wt of substance (col. 7)}}{\text{wt of sample (col. 2)}} \times 100$$

C. *Determination of the Equivalent Weight of an Unknown.* Before you use the analytical balance, your instructor will demonstrate the proper procedure in using this important and expensive piece of equipment. Practice weighing sand as if it were your unknown, as described in "The Use and Care of Balances" on pages xiii-xiv.

On the analytical balance, weigh two separate samples of your unknown solid acid on accurately weighed glassine papers. Use an amount of solid acid in the range of 150-200 mg and record all weighings to the fourth decimal place. Transfer each sample quantitatively to a 125- or 250-ml Erlenmeyer flask using a camel-hair brush or a stream of distilled water (wash bottle). Dissolve the sample in 25-40 ml of distilled water. Add 3 drops of 1% phenolphthalein indicator and titrate to a faint pink color with standard NaOH solution. Record the data on the laboratory report sheet and calculate the equivalent weight of your unknown acid to the first decimal place.

At the end point of the above titration, the milliequivalents ($ml_b \times N_b$) of the NaOH required to neutralize a known weight of your unknown acid will exactly equal the milliequivalents of this acid. The milliequivalent weight of your acid can be calculated from the equation: meq of acid $\times$ meq wt of acid = wt of acid. To do this, substitute ($ml_b \times N_b$) for "meq of acid" and rearrange the equation into the following form:

$$\text{meq wt of acid} = \frac{\text{weight of acid}}{ml_b \times N_b}$$

provided the sample weight of the acid and the milliequivalent weight of the acid are expressed in the *same units* of mass (mg or g). Remember that the eq wt = $1000 \times$ meq wt.

Reagents, Materials, and Equipment are listed on page 275.

LABORATORY REPORT

Name_____ Section_____Date_____

Experiment 8. TOTAL ACIDITY

A. *Practice in Titration*

Normality of $H_2SO_4 =$ _____N   Normality of NaOH = _____N

	Trial 1	*Trial 2*	*Trial 3*
Ml of H_2SO_4	_____	_____	_____
Ml of NaOH	_____	_____	_____
Meq of H_2SO_4 (ml $\times$ N)	_____	_____	_____
Meq of NaOH (ml $\times$ N)	_____	_____	_____

B. *Total Acidity or Alkalinity of Various Materials*

1. Material analyzed	2. Wt of sample	3. Volume reagent used	4. N of reagent	5. Meq. of reagent used	6. Meq wt of subst. sought (g)	7. Wt of substance sought	8. % of substance found
Household Ammonia*							

*Household ammonia (density = 1 g/ml) was diluted 1 ml to 10 ml (reagent shelf) and 5 ml of this diluted sample was titrated. What weight of undiluted household ammonia was titrated? Enter this weight in column 2.

C. *Determination of the Equivalent Weight of an Unknown Acid.*

Sample No. _____   1   2

Wt of paper + sample	_____	_____
Wt of paper alone	_____	_____
Wt of sample	_____	_____
Ml of NaOH	_____	_____
Normality of NaOH _____ N		
Meq of NaOH	_____	_____
Eq wt of unknown in grams	_____	_____

QUESTIONS AND PROBLEMS

Name_____ Section_____ Date_____

Experiment 8. TOTAL ACIDITY

NOTE: Show the mathematical steps required to solve each problem and the units of volume, weight, etc., associated with the numerical values in the data and answers.

1. Define each of the following terms:

   a. 1 M solution of a solute. _____

   _____

   b. 1 N solution of an acid. _____

   _____

   c. 1 N solution of a base. _____

   _____

   d. Gram-equivalent weight of an acid. _____

   _____

   e. Gram-equivalent weight of a base. _____

   _____

2. Calculate the amount of pure acid or base required to prepare the following solutions:

   a. 2 liters of 0.1 N KOH   b. 5 liters of 1.0 N H_2SO_4

3. If 22.70 ml of NaOH neutralizes 25.0 ml of 0.100 N HCl, what is the normality of the NaOH solution?
   Ans. 0.110 N

4. Complete the following table showing the number of milliequivalents, the milliequivalent weight, and the weight of each substance given below when a solution of the substance is titrated with the reagent indicated.

Substance titrated	Volume and Normality of reagent used	Milli- equivalents	Meq wt grams	Weight grams
a. lactic acid	20.1 ml of 0.105 N NaOH	_____	_____	_____
b. citric acid	42.2 ml of 0.105 N NaOH	_____	_____	_____
c. sodium bicarbonate	28.5 ml of 0.095 N H_2SO_4	_____	_____	_____
d. pot. acid tartrate	16.5 ml of 0.105 N NaOH	_____	_____	_____

5. If 26.0 ml of 0.110 N NaOH are required to neutralize 0.329 g of an acid, what is the equivalent weight of the acid?   Ans. 115 g/eq

6. How many ml of 0.20 M H_2SO_4 are needed to neutralize 34.0 ml of 0.10 M NaOH?   Ans. 8.5 ml

7. To neutralize the HCl in 50 ml (50 g) of gastric juice, 38.5 ml of 0.100 N NaOH are required. What is the percentage of HCl in the gastric juice?   Ans. 0.28% HCl

8. If 5 ml (5g) of household ammonia are diluted to 100 ml with distilled water and a 25-ml aliquot requires 29.6 ml of 0.107 N acid for titration, what percentage of ammonia (NH_3) is present in the household ammonia?   Ans. 4.31% NH_3

Carbohydrates

9

General Color Test for Carbohydrates

INTRODUCTION

Carbohydrates are dehydrated when treated with strong mineral acids under nonoxidizing conditions. This dehydration occurs in 4-6 N acid at 100°C or in concentrated acids at lower temperatures. If the carbohydrate is an oligosaccharide (e.g., disaccharide, trisaccharide, etc.) or a polysaccharide, hydrolysis of the carbohydrate acetal linkage occurs simultaneously with the dehydration reaction.

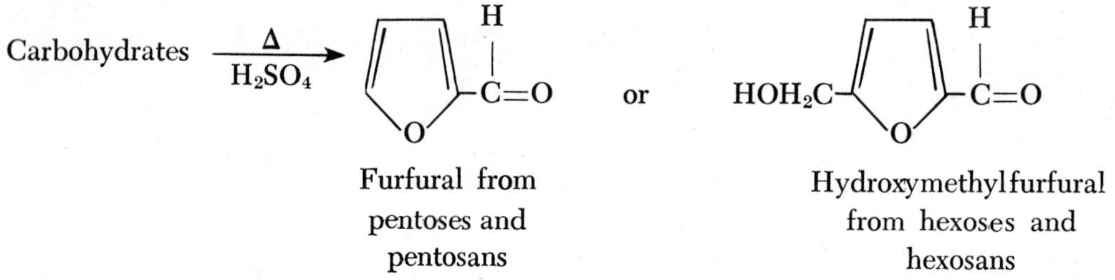

Carbohydrates $\xrightarrow[\text{H}_2\text{SO}_4]{\Delta}$

Furfural from
pentoses and
pentosans

or

Hydroxymethylfurfural
from hexoses and
hexosans

Pentoses and pentosans are converted to furfural by concentrated H_2SO_4 or HCl, while hexoses and hexosans form primarily hydroxymethylfurfural. These cyclic aldehyes, in turn, will react with α-naphthol or anthrone to form a mixture of colored condensation products, which are the basis of the Molisch and anthrone tests for carbohydrates. Under appropriate conditions the anthrone reaction can be used as a quantitative assay for carbohydrates.

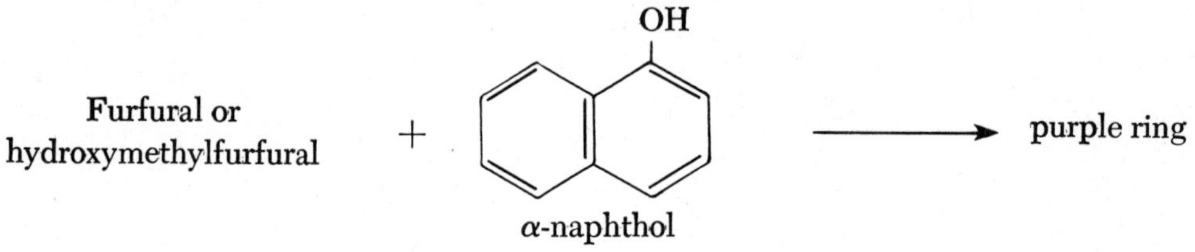

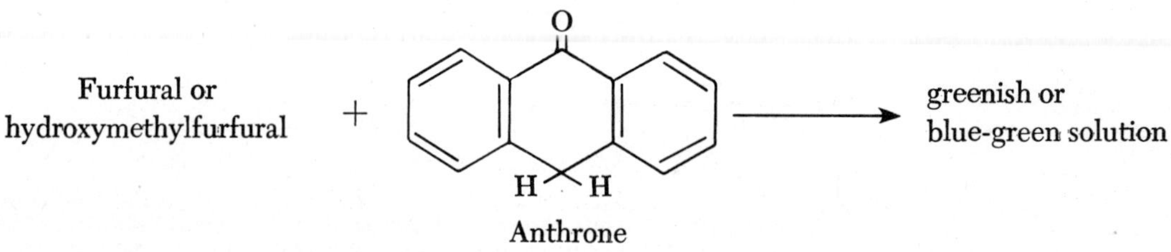

Furfural or
hydroxymethylfurfural

+

α-naphthol

⟶ purple ring

Furfural or
hydroxymethylfurfural

+

Anthrone

⟶ greenish or
blue-green solution

Furfural and hydroxymethylfurfural react with phenolic compounds and aromatic amines to yield colored products. However, in spite of identical reagents and reaction conditions, the different sugars will dehydrate at different rates to form the furfurals, which may produce a variety of colored condensation products with phenolic reagents. (Note the rate of color formation of the different carbohydrates used in the Molisch test in part A.) These different rates of dehydration and the various colors formed by the same reagent is the basis of qualitative tests for certain sugars or types of sugars (e.g., the resorcinol test for ketoses and the orcinol test for pentoses in Expt. 11).

The instructor may prefer to demonstrate this particular experiment to the entire class rather than have each student perform it individually.

PROCEDURE

A. *Molisch Test.* This test is given by all soluble carbohydrates, and a similar color is developed at the surface of insoluble carbohydrates under appropriate conditions.

To three separate test tubes add 10 drops of 0.5% solutions of glucose, sucrose, and starch, respectively, and dilute each sugar solution with about 2 ml of water. Add 2 drops of α-naphthol solution to each tube and mix. Incline the test tube and slowly and carefully add about 3 ml of concentrated sulfuric acid* down the side of the tube to form a layer below the sugar solution. A *purple ring* at the interface is indicative of a carbohydrate.

B. *Anthrone Test.* To three separate test tubes add 1 drop of 0.5% solutions of glucose, sucrose, and starch, respectively, and dilute each solution with 1 ml of water. Carefully add 3 ml of anthrone reagent (**CAUTION:** made with conc. H_2SO_4), mix thoroughly by swirling, heat in a boiling water bath for 3 minutes, cool and observe the color formed. The mixture should remain clear, and should assume a greenish or blue-green color of an intensity depending on the amount of carbohydrate present. If instead it becomes cloudy or turbid, this means that too much water has been mixed with the reagent. In this case add a little more reagent. If the test solution is too dark or opaque repeat the test with a more dilute solution of carbohydrate.

Reagents and Materials are listed on page 276.

*Special dispensing burets, which can be attached to the concentrated nitric and sulfuric acid bottles, are available for the student's convenience and safety. See Appendix, p. 268.

LABORATORY REPORT

Name _____ Section _____ Date _____

Experiment 9. GENERAL COLOR TESTS FOR CARBOHYDRATES

A. *Molisch Test*

Compare the purple ring formed by the carbohydrates tested in regard to the rate of formation, intensity of color, and anything else you might observe.

Glucose _____

Sucrose _____

Starch _____

B. *Anthrone Test*

Compare the colors formed by the three carbohydrates tested.

_____

QUESTIONS

1. Assuming 15 drops per ml, what weight of glucose is present in the volumes of 0.5% glucose solution given below?

    a. g glucose/100 ml 0.5% glucose =_____g;   g glucose/1 ml =_____g

    b. mg glucose/15 drops =_____mg;   mg glucose/1 drop =_____mg
    Note the small amount of glucose (or other carbohydrate) per 1 drop.

2. a. What are the products of hydrolysis from the following carbohydrates?

Carbohydrate	Hydrolysis Product(s)
Starch	_____
Sucrose	_____
Lactose	_____
Glycogen	_____
Xylan	_____

    b. Which of the above dehydrate to furfural?

    _____

3. What is a glycoprotein or mucoprotein and how would they respond to the Molisch or anthrone test?

4. a. Name the disaccharide in milk. _____

    b. Milk contains about 5% of this disaccharide. How would you proceed in proving its presence in milk by the anthrone test? (Recall question 1 and the fact that hot, concentrated H_2SO_4 also reacts readily with other organic constituents of milk.)

5. Do all carbohydrates react with H_2SO_4 at the same rate and produce the same color with the color forming reagents? Explain briefly, including your own observations on the carbohydrates tested in this experiment.

Carbohydrate Tests Involving Oxidation of Sugars

INTRODUCTION

The free or potentially-free aldehydic or ketonic groups of monosaccharides and certain disaccharides are readily oxidized by mildly alkaline solutions of cupric ions, such as Benedict's reagent (copper sulfate, sodium carbonate, and sodium citrate). These carbohydrates are called *reducing sugars* because they reduce cupric ions to form cuprous oxide, Cu_2O, a yellow to brick-red precipitate. The aldoses are oxidized to *aldonic acids* (i.e., glucose forms gluconic acid), as well as other products. The ketoses yield a mixture of lower molecular weight oxidation products. The alkalinity of the Benedict's solution (pH 10.5) is due to the hydrolysis of sodium carbonate. Were it not for the presence of the sodium citrate in this alkaline solution, the cupric ions would precipitate as basic cupric carbonate in which a part of the CO_3^{2-} ion in the basic $CuCO_3$ is replaced by OH^- ions. Instead, the Cu^{++} ions form a soluble, stable complex ion with the citrate.

Barfoed's reagent (copper acetate in 1% acetic acid) is a slightly acidic solution (pH 4.6) of Cu^{++} and is a weaker oxidizing agent than the mildly alkaline Benedict's solution. On the other hand, monosaccharides are stronger reducing agents toward Cu^{++} than the disaccharides. Based on these chemical properties, the Barfoed's reagent distinguishes between monosaccharides and disaccharides, because the former will reduce Cu^{++} to Cu_2O in 3½ minutes while the disaccharides will not. However, when disaccharides are classified as being *reducing* or *non-reducing* disaccharides, the classification is based on whether they will or will not reduce Cu^{++} in an *alkaline* solution (e.g., Benedict's solution) and *not* how they react with Barfoed's solution.

When a solution of an aldose is heated with nitric acid, an *aldaric acid* results from strong oxidation. Both the aldehydic and hydroxymethyl groups at the ends of the aldose carbon chain are oxidized to form carboxyl groups. For example, glucose yields glucaric acid, while galactose forms galactaric or mucic acid. The *insolubility* of mucic acid in water, in contrast to all other aldaric acids, is the basis for the *mucic acid test* for galactose, either free or as a part of oligo- or polysaccharides.

PROCEDURE

Unknown Carbohydrate. Your unknown will be a 2% solution of one sugar or a mixture of two sugars (each 2%) from the list given in Experiment 12 "Summary of Sugar Tests." Use 10 ml of the solution to set up the mucic acid test as instructed in part C. Dilute the remainder of the 2% unknown solution with an equal volume of distilled water to make a 1% solution for the Benedict's and Barfoed's tests in this experiment as well as for the tests in Experiment 11. Do the tests for the unknown and for the known sugars at the same time. Refrigerate unused sugar unknown between lab periods.

A. *Benedict's Test for Reducing Sugars.* Perform the test on 1% solutions of glucose, xylose, fructose, sucrose, lactose, starch, and your unknown. Add 5 ml of Benedict's reagent and 1-2 ml of the carbohydrates to separate test tubes and *shake* each tube thoroughly. Place all tubes in a *boiling* water bath at the same time. (Use a 600 or 800-ml beaker one-third filled with distilled water plus a few boiling stones to prevent bumping.) Heat in the water bath for 5-6 minutes and observe what happens. On the report sheet describe the changes in the color of the solutions, in the transparencies, and in the formation and color of any precipitates. This can be done better if the tubes are allowed to stand in the test tube rack for a few minutes. Compare the results in each case with that obtained with glucose.

Add 4-5 drops 3 M HCl to 5 ml of 1% sucrose solution and heat in the boiling water bath for 5 minutes. Treat 5 ml of 1% starch solution in the same way, but extend the heating period to 25-30 minutes. Apply the Benedict's test to 1-2 ml of each solution in the same manner as described earlier. Compare the results with those obtained *without* the acid treatment. Keep the remainder of the two acid-hydrolized solutions for the Barfoed's test in part B.

B. *Barfoed's Test.* To 5 ml of Barfoed's reagent in separate test tubes add 5 ml of 1% solutions of glucose, xylose, fructose, sucrose, maltose, starch, and your unknown, respectively. *Shake* the contents of each tube well, and place *all* of the tubes at the *same time* in an *actively* boiling water bath. Heat for 3½ minutes after the water starts boiling again. Timing is important, since a positive test can be obtained with disaccharides if they are heated more than 3½ minutes. During this period observe the tubes closely and note any change in color or clarity of the solutions. Record the time if a reaction (turbidity) occurs. For purposes of comparison, it is helpful to make a control tube with 5 ml of water in place of the sugar solution and heat at the same time.

Using the remainder of the acid-hydrolyzed 1% sucrose and 1% starch solutions from part A, add 5 ml of Barfoed's reagent to each of the two tubes and heat in the actively-boiling water bath for 3½ minutes. Compare the results from these acid-treated solutions with those obtained from the Barfoed's test on untreated sucrose and starch.

C. *Mucic Acid Test.* **NOTE:** In this experiment it is necessary to use *concentrated* HNO_3,* a dangerously corrosive liquid. Take care that you do not allow it to spatter or to drip down on the outside of various containers from which it has been poured. Any drops that are spilled on the tables, desk tops, or floor must be wiped up at once with a wet cloth or sponge. Nitric acid on the skin or clothing must be washed off immediately with large amounts of water, rubbing the spot vigorously to work out all the acid. These remarks apply equally to concentrated H_2SO_4, NaOH, or other corrosive liquids, wherever they are encountered in laboratory work. It is essential that each student use *extreme care* in handling these materials to avoid personal injury and damage to equipment.

Use 10 ml of 2% solutions of galactose and your unknown in separate porcelain evaporating dishes. Add 5 ml *concentrated* nitric acid* to each dish. Support each dish on a 250-ml beaker that is two-thirds filled with distilled water. Evaporate *(under hood)* the contents of the dishes on the boiling water baths to about a volume of 2 ml. (Evaporation can be done directly on a steam bath under a hood.) Evaporation takes about 45 minutes; do not allow water baths to go dry during this time. Remove the hot dishes with tongs and allow to cool. Cover the dishes with watch glasses and set aside in your locker until the next laboratory period. Be sure your dishes are properly identified.

Add 5 ml of distilled water to each dish and mix well with a stirring rod, making every effort to possibly dissolve any residue from the oxidation of your unknown. Note the insoluble mucic acid crystals from the oxidation of galactose. Compare your unknown results with those obtained from the known galactose. Sometimes the comparison can be made better if the contents of the dishes are poured into clean test tubes.

NOTE: In addition to completing the Laboratory Report for this experiment, also make entries in the "Summary of Sugar Tests" on page 83 for the Benedict's, Barfoed's and Mucic Acid tests for *all* single sugars and sugar mixtures listed in the table.

Reagents and Materials are listed on page 276.

---

*Special dispensing burets, which can be attached to the concentrated nitric and sulfuric acid bottles, are available for the student's convenience and safety. See Appendix, p. 268.

LABORATORY REPORT

Name____________________ Section__________ Date__________

Experiment 10. CARBOHYDRATE TESTS INVOLVING OXIDATION OF SUGARS

A. ***Benedict's Test***—A mild alkaline oxidation.

Describe the changes in the color and transparencies of the solutions, and in the formation and color of the precipitates. Compare the intensity of the reaction in each case with that given by glucose.

Glucose ____________________

Xylose ____________________

Fructose ____________________

Sucrose ____________________

Lactose ____________________

Starch ____________________

Unknown ____________________

What was the result of the Benedict's test on sucrose and starch solutions *after* they had been heated with dilute HCl? Compare with results obtained before acid treatment.

Sucrose ____________________

Starch ____________________

B. **Barfoed's Test**—A mild acidic oxidation.

Record the time required for a solution to become cloudy. If solution remains clear after heating 3½ minutes, indicate *No Reaction.*

Sugar tested	Minutes to form ppt. (if any)	Indicate if sugar gave + or − test
Glucose	_____	_____
Xylose	_____	_____
Fructose	_____	_____
Sucrose	_____	_____
Maltose	_____	_____
Starch	_____	_____
Unknown	_____	_____

What were the results of the Barfoed's test on the remainder of the sucrose and starch solutions that had been heated with dilute HCl in part A? Compare with results of Barfoed's test on untreated sucrose and starch.

Sucrose _____

_____

Starch _____

_____

What product(s) is (are) obtained from the complete acid hydrolysis

Of sucrose?_____

Of starch? _____

C. **Mucic Acid Test for Galactose**—A strong acidic oxidation.

Describe the appearance of the mucic acid crystals._____

_____

Result with unknown _____

Write an equation for the oxidation showing *structural* formulas of galactose and the mucic acid.

QUESTIONS

Name_____ Section _____ Date _____

Experiment 10. CARBOHYDRATE TESTS INVOLVING OXIDATION OF SUGARS

1. When a sugar gives a positive test with Benedict's reagent, which reactant undergoes oxidation and which undergoes reduction? Specify the oxidizing and reducing agents involved.

    a. Reactant oxidized _____ ; oxidizing agent_____

    b. Reactant reduced_____ ; reducing agent _____

2. Give the name and structure of the functional *group* in aldoses and ketoses responsible for a positive Benedict's test.

    Aldoses Name:_____   Structure:

    Ketoses Name:_____   Structure:

3. What is meant by the term *reducing sugar?*

   _____

4. In an aqueous solution, glucose occurs primarily in the α- and β-forms of the *pyranose* hemiacetal ring structure that does not have the functional group required for a positive Benedict's test. Why, then, do all of the glucose molecules react with Cu^{++}?

5. The formation of a red-orange precipitate is the *principal* evidence of a positive test. What other color change occurs simultaneously in a positive test?

   _____

6. a. Cite examples from your tests that show that a negative Benedict's test is not always indicative of the absence of a sugar or other carbohydrate.

   _____

   b. Lactose is a disaccharide with an acetal linkage of $\beta,1(\text{gal})\rightarrow4(\text{glu})$ and sucrose is a disaccharide with an acetal linkage of $\alpha,1(\text{glu})\rightarrow\beta,2(\text{fru})$. Explain why one is a reducing sugar and the other one is not. (Refer to your answer to question 4 and the structures of lactose and sucrose.)

7. What happens to any oligo- or polysaccharide initially when they are heated with a *dilute* solution of a strong acid?

   _____

8. Indicate with a + or − whether each of the following classes of carbohydrates reacts positively or negatively to Barfoed's reagent.

_____ aldohexoses _____ reducing disaccharides _____ monosaccharides

_____ ketohexoses _____ nonreducing disaccharides _____ polysaccharides

What is meant by a "reducing disaccharide" versus a "nonreducing disaccharide"?

_____

9. Compare Benedict's and Barfoed's solutions as follows:

  a. Chemical composition: Benedict's _____

  Barfoed's _____

  b. Acidity or alkalinity: Benedict's _____

  Barfoed's _____

  c. Relative rates of reaction with reducing sugars. _____

  _____

10. Why is it important that heating time be no longer than 3½ minutes?_____

  _____

11. Of what use is this test in correctly identifying an unknown sugar? Explain. _____

  _____

12. a. Name the reaction product of glucose. $\xrightarrow[\text{HNO}_3]{\text{evap.}}$ _____

  b. How could you distinguish the product in a. from mucic acid?

  _____

13. Name the products formed upon heating a 2% lactose solution with conc. HNO_3.

  lactose $\xrightarrow[\text{heat}]{\text{H}_2\text{O/H}^+}$ $\xrightarrow[\text{HNO}_3]{\text{evap.}}$

14. Raffinose is a naturally occurring trisaccharide in which the three monosaccharide units are linked as follows: $\alpha,1(\text{gal}) \rightarrow 6(\text{glu})\alpha,1 \rightarrow \beta,2(\text{fru})$.

  a. Would raffinose give a positive mucic acid test? Explain. _____

  _____

  b. If an enzyme would selectively hydrolyze only the $\alpha,1(\text{gal}) \rightarrow 6(\text{glu})$ linkage of raffinose, what

  products would be formed?_____

  c. Is raffinose a reducing or a nonreducing sugar? _____

15. Would galacuronic acid give a positive mucic acid test? Explain._____

  _____

Color Tests for Ketoses and Pentoses

INTRODUCTION

Heating with HCl dehydrates hexoses to hydroxymethylfurfural (see Expt. 9), levulinic acid, and other products. Ketohexoses yield larger amounts of hydroxymethylfurfural, and at a faster rate, than do the aldohexoses. These differences are the basis of the Seliwanoff test for ketoses, in which *resorcinol* forms a red condensation product with hydroxymethylfurfural. Using dilute sugar solutions (less than 2 mg of a sugar in the test solution) and avoiding prolonged heating, one can readily distinguish between ketoses and aldohexoses, which produce a light yellow to faintly pink color under appropriate test conditions. As pointed out in Experiment 9, oligosaccharides (disaccharides, etc.) and polysaccharides readily hydrolyze in hot acid solutions, so that any of these carbohydrates containing a ketose unit will respond positively to the Seliwanoff resorcinol test. Pentoses are dehydrated to furfural, which forms a greenish color with resorcinol, and students must remember this possible interference when observing the results from their unknown.

The furfural produced by the reaction of hot dilute HCl on pentoses forms a bright red color with *aniline acetate* in a test paper held over the mouth of the reaction flask. The hydroxymethylfurfural derived from the hexoses does not interfere, because it is much less volatile than furfural. Hence, the critical feature of this test is to allow only the *vapors* from the hot mixture to come in contact with the aniline acetate. It should also be noted that the hot acid solution will hydrolyze polysaccharides into their component monosaccharides, so that pentosans will also respond to this test. The hulls of the seeds of cereal grains are rich in pentosans (hemicelluloses). Xylan is a hemicellulose present in wood associated with the cellulose component. The acid hydrolysis of wood is a good source of xylose, commonly known as wood sugar.

Bial's *orcinol* test is a simple, rapid qualitative test for pentoses. Orcinol (5-methylresorcinol) gives a green color with the furfural from dehydrated pentoses (or pentosans). As in the case of the Seliwanoffs resorcinol test, *dilute* solutions of the sugar must be used. Ketoses yield a yellow-green color, which can interfere, while the light yellow-brown color from aldoses interferes to a lesser degree. Hence, the orcinol test lacks the selectivity of the aniline acetate test, especially in the presence of hexoses. Nevertheless, the orcinol reaction is used in the quantitative assay for pentoses in the absence of interfering substances as, for example, in the analysis of ribonucleic acids (see Expt. 39).

PROCEDURE

A. *Seliwanoff's Resorcinol Test for Ketoses.* Less than 2 mg of sugars must be used in this test. Place 2-3 drops of 1% solutions of fructose, glucose, sucrose, sorbose, xylose, and your unknown in separate test tubes. Add 5 ml of the resorcinol reagent (which contains the HCl) to each test tube. Set all tubes in a beaker two-thirds filled with boiling water. Record the color you observe in each tube at the end of 1 minute and after 4 minutes. The red color from ketoses after 4 minutes constitutes a positive test. Show tests to the instructor.

B. *Aniline Acetate Test for Pentoses.* Place 5 ml of 1% xylose and 20 ml water in a 250-ml Erlenmeyer flask, add 20 ml of concentrated HCl, and boil the resulting solution gently for about one minute on a wire gauze *under the hood.* Discontinue the heating, hold a filter paper moistened with a few drops of aniline acetate over the mouth of the flask, and observe any color changes on the paper. Do *not* allow the paper to rest on the mouth of the flask (*hold* the paper); otherwise false results may be obtained. Repeat with separate portions of glucose, fructose, lactose, starch, and your unknown. Also test a 1 to 2-g portion of wheat germ or shredded wheat. Complete the table on the report sheet.

C. *Bial's Orcinol Test for Pentoses (Optional).* (**NOTE:** The reagent *ordinarily* is a solution of 0.1% orcinol in concentrated HCl containing 0.1% $FeCl_3 \cdot 6H_2O$. Equal volumes of the reagent and the dilute sugar are mixed. In this experiment *one drop* of a 1% sugar solution is being used, so that an equal volume of water has already been added to the orcinol reagent on the shelf.)

Into *separate* test tubes, place one drop of a 1% solution of xylose, glucose, fructose, lactose, starch, and your unknown. Also place several particles of shredded wheat into one tube. Add 3-4 ml of orcinol reagent to each tube and heat the tubes in a boiling water bath for 3-5 minutes. The characteristic green color in the xylose tube is a positive test for pentoses. The various colors of the other carbohydrates constitute a negative test. Record colors and indicate whether test is + or −. Show results to your instructor.

NOTE: Complete the table on page 83, "Summary of Sugar Tests," for all sugars and sugar mixtures listed.

Reagents and Materials are listed on page 277.

Name _____ Section _____ Date _____

Experiment 11. COLOR TESTS FOR KETOSES AND PENTOSES

A. *Resorcinol Test for Ketoses*

Complete the table for results on the dilute sugar solutions listed.

Name of sugar	Color of solution after	
	1 minute	4 minutes
Fructose	_____	_____
Glucose	_____	_____
Sucrose	_____	_____
Sorbose	_____	_____
Xylose	_____	_____
Unknown	_____	_____

B. *Aniline Acetate Test for Pentoses and C. Bial's Orcinol Test*

In the second column of the table, list in the blank spaces the carbohydrates you tested which correspond to the appropriate classification in the first column. Indicate color obtained and a + or −.

Carbohydrate tested		Give color. Indicate + or −.	
Classification	Name	Aniline acetate	Bial's orcinol
Aldopentose			
Aldohexose			
Ketohexose			
Disaccharide			
Glucosan			
Pentosan	Shredded Wheat		
Unknown	. . . . . .		

Name _____ Section _____ Date _____

Experiment 11. COLOR TESTS FOR KETOSES AND PENTOSES

A. *Resorcinol Test for Ketoses*

1. Like ketohexoses, the aldohexoses can dehydrate to hydroxymethylfurfural, which condenses with resorcinol to give a pink to red color. List several controlling factors that prevent aldoses from interfering with the Seliwanoff test for ketoses.

   _____

   _____

2. a. Draw the
      structure
      of
      resorcinol

   b. L-sorbose is the C-5 epimer of D-fructose. Draw the open-chain structure of L-sorbose in the space to the right.

3. Ketohexoses are monosaccharides. Why will sucrose (a disaccharide), raffinose (a trisaccharide), and inulin (a fructosan polysaccharide) give a positive Seliwanoff test?

   _____

   _____

4. Compare the milligrams of sugar used in the Benedict's test with that used in the Seliwanoff test. Assume 20 drops per ml.
   a. Benedict's test: 2 ml 1% sugar. Show work.

      _____ mg sugar
   b. Seliwanoff test: 3 drops 1% sugar. Show work.

      _____ mg sugar
   c. Why is it necessary to use such small amounts of sugar in the Seliwanoff test?

   _____

   _____

B. *Aniline Acetate and Bial's Orcinol Test*

1. Describe in detail the *appearance of the paper* when the aniline acetate test is (a) positive, (b) negative. Is any red color produced when hexoses are tested?

   Positive test: _____

   Negative test: _____

   Color with hexoses: _____

2. Both furfural from pentoses and hydroxymethylfurfural from hexoses will give a red color with aniline acetate. Why don't hexoses interfere in the aniline acetate test when properly done?

_____

_____

3. Shredded wheat contains mainly starch, a hexosan. Why does it give a good positive test with aniline acetate?

_____

_____

4. Which test, aniline acetate or orcinol, is more specific in testing for pentoses? Explain.

_____

_____

5. Would the following substances give a positive aniline acetate test? Explain in each case.

   a. Ribonucleic acids (RNA)_____

   b. Refined wheat flour _____

   c. Wood pulp _____

6. In order to determine the presence or absence of a pentose in a mixture of sugars, why is a *dilute* solution of the sugars used in the orcinol test, especially in the presence of a ketose?

_____

_____

7. Write the structures of the following:
   a. Orcinol   b. Furfural   c. Hydroxymethylfurfural
   (5-methylresorcinol)

Summary of Sugar Tests. Identification of Unknown Sugars

Name_____ Section _____ Date _____

A. *Summary of Carbohydrate Tests*

Indicate with a + or — whether the sugars listed below give a positive or negative test with each of the reagents. Fill in *all* spaces, even if the sugar may not have been included in a particular test in a previous experiment.

Sugar tested	Benedict's	Barfoed's	Resorcinol	Mucic acid	Aniline acetate	Orcinol
Glucose						
Fructose						
Galactose						
Sucrose						
Lactose						
Maltose						
Xylose						
Fructose + Xylose						
Galactose + Xylose						
Galactose + Fructose						
Sucrose + Lactose						
Sucrose + Maltose						
Unknown						

B. *Identification of an Unknown*

Your unknown is one of the sugars or a mixture of two sugars in the list given below. Based on the results from your sugar tests as shown in the table in part A, identify your unknown.

Unknown Sample No. _____   My unknown is _____

List of unknowns:

1. glucose	7. xylose
2. fructose	8. fructose + xylose
3. galactose	9. galactose + xylose
4. sucrose	10. galactose + fructose
5. lactose	11. sucrose + lactose
6. maltose	12. sucrose + maltose

QUESTIONS

Name _____ Section _____ Date _____

Experiment 12.   SUMMARY OF SUGAR TESTS. IDENTIFICATION OF UNKNOWN SUGARS

1. In considering the various single sugars and mixtures of sugars in the summary table:

   a. Which is the only sugar in the list that gives a *negative* Benedict's test?_____

   b. Does a *positive* Barfoed's test indicate that you do not have a disaccharide in your unknown? Explain.

   _____

   c. What does a *negative* Barfoed's test tell you?

   _____

   d. If you carelessly heated the Barfoed's test for more than 3½ minutes and obtained a false positive test, what sugars would you incorrectly report for the following unknowns? Restrict your answer to the list of unknowns given on page 84.

   lactose: _____   sucrose + lactose: _____

   maltose: _____   sucrose + maltose: _____

2. a. Which sugars would be missed or incorrectly reported if the mucic acid test should have been positive and you reported it as being negative?

   _____

   b. Select a single sugar and a mixture of two sugars from the unknown list and indicate how you would have incorrectly reported them if you had *missed* a positive mucic acid test.

   1. Single sugar: _____

   2. Mixture of two: _____

3. a. If your unknown were sucrose and maltose (no. 12 on the unknown list), how did you know that you had these two sugars and not simple fructose alone?

   _____

   b. Supposing you were given an unknown mixture of fructose and sucrose (not on the list): Which sugar would you find to be *definitely* present? Explain.

   _____

   Which sugar would you be unable to prove to be definitely present? Explain.

   _____

4. If your unknown was a mixture of fructose and xylose,

    a. In which color test is xylose likely to interfere?_____

    b. Why is the aniline acetate test for xylose more reliable than the orcinol test?

_____

_____

Detection of Glucose in Urine

INTRODUCTION

The detection of glucose in urine is of particular importance in the control of diabetes mellitus. The two tests illustrated in this experiment involve (1) the reduction of Cu^{++} to cuprous oxide by glucose, and (2) the oxidation of glucose by molecular oxygen in the presence of two enzymes and a chromogen to produce a color change on the test paper.

The CLINITEST reagent tablets contain copper sulfate pentahydrate, sodium carbonate, sodium hydroxide, and citric acid. When the tablet is dropped in the urine and water mixture, the solution becomes very hot from the heat of solution of NaOH. Some of the citric acid reacts with the dissolved NaOH to form sodium citrate. The citric acid also reacts with the Na_2CO_3 to produce an effervescence, which thoroughly mixes the test solution. These reactions result in the formation of a reagent mixture, which is essentially identical with Benedict's solution (Expt. 10). The color of the reaction mixture after the test period is proportional to the cuprous oxide formed by the reducing substances in urine, with glucose being the principal reducing material in the case of glucosuria. The color chart supplied with the test tablets covers a range of 0 to 2% glucose.

The two types of test papers commonly used for the detection of glucose in urine are CLINISTIX (Ames), a plastic strip with a tuft of paper attached to one end, and TES-TAPE (Lily), a roll of test paper in a dispenser similar to the Hydrion paper used in Experiment 6. Both papers are impregnated with a buffered mixture of two enzymes, glucose oxidase and a peroxidase, and with a chromogen, o-tolidine. An additional dye, tartrazine, is present in TES-TAPE to increase the range of glucose estimation. When the test paper is wetted with urine containing glucose, the following reactions take place:

$$\text{D-glucose} + H_2O + O_2\,(\text{air}) \xrightarrow{\quad\text{glucose oxidase}\quad} \text{D-gluconic acid} + H_2O_2 \qquad (1)$$

$$H_2O_2 \xrightarrow{\quad\text{peroxidase}\quad} H_2O + [O]\ \text{active oxygen} \qquad (2)$$

$$\text{o-tolidine (reduced)} + [O] \xrightarrow{\qquad\qquad} \begin{array}{c}\text{o-tolidine (oxidized)}\\ \text{(blue color)}\end{array} \qquad (3)$$

The glucose oxidase reaction is specific for glucose. If galactose, lactose, or fructose are present in the urine specimen, they will give a positive result with the CLINITEST tablets, but a negative result with the test papers, CLINISTIX and TES-TAPE. Excessive amounts of reducing substances like creatinine, uric acid, or glucuronic acid can also give a false positive test with CLINITEST tablets.

Ingestion of large amounts of vitamin C (e.g., more than 0.5 g daily) or the administration of parenteral preparations in which the ascorbic acid is present as a reducing agent will interfere with both the Cu^{++} reduction and the glucose oxidase methods. When using the CLINITEST tablets, the vitamin C in urine reduces Cu^{++} to cuprous oxide, giving either a false positive or high positive test. With the use of CLINISTIX or TES-TAPE, vitamin C may act as an antioxidant inhibiting reaction (1) above, or as a reducing agent preventing or reversing reaction (3), giving either a false negative or a low positive test if glucose is present in the urine specimen. A person who is monitoring his insulin dosage to control a diabetic condition must avoid taking large amounts of vitamin C to combat the common cold, a controversial procedure recommended by Linus Pauling, a Nobel prize laureate. As little as 0.1 mg ascorbic acid per 1 ml urine can affect the glucose test results.

The *metabolites* from certain pathological conditions or from the use of certain drugs are reducing substances that have been shown to interfere with the color formation of reaction (3) of the glucose oxidase test, as well as giving a false or high positive test with CLINITEST tablets. Homogentisic acid from faulty phenylalanine metabolism (alcaptonuria), 3,4-dihydroxyphenylacetic acid from L-dopa therapy for Parkinson's disease, and gentisic acid from large doses of aspirin are examples of interfering reducing substances. Aspirin itself does not interfere, but one of its metabolites, gentisic acid, does. A more detailed account of interfering metabolites and their possible effect on the test methods is found in the literature. (See Reference.) Although the above methods for the detection of glucose in urine are widely and successfully used, one must always be alert for potential interferences where medications and chemotherapy are used.

In this experiment the urine specimens are fortified with glucose at several levels so that the student can observe the appearance of a positive glucose test. In addition, one sample will contain vitamin C, so that you can note its effect.

PROCEDURE

A. *Detection of Glucose with CLINITEST Reagent Tablets.* The reagent tablets normally have a white color with blue spots due to $CuSO_4 \cdot 5H_2O$. Tablets that have become intensely blue should not be used because they have absorbed moisture (and CO_2) and will no longer give proper heating and color. No external heating is required, since the tablets generate their own heat during the test. Handle tablets carefully because they contain caustic soda. Keep bottle *tightly* closed when not in use.

The student is encouraged to bring his own urine specimen to test for glucose. For cleanliness and for prevention of contamination, dropper bottles containing urine, 1.0% glucose, vitamin C, and distilled water will be available. Using *one* disposable test tube (13×100 mm) perform the tests given below. Thoroughly rinse the tube after each test, reuse it, and *discard* it when you are finished.

A — 5 drops urine + 10 drops water
B — 5 drops urine + 4 drops 1.0% glucose + 6 drops water
C — 5 drops urine + 10 drops 1.0% glucose
D — 5 drops urine + 10 drops 0.5% vitamin C.

Using *one sample at a time,* proceed as follows: (also see color chart)

1. Drop one CLINITEST reagent tablet into the test tube and watch while the complete reaction takes place. Do *not* shake the test tube during the reaction (see NOTE below), nor for 15 seconds after boiling has stopped.

2. After the 15-second waiting period, gently shake the test tube and compare the color of the mixture with the color chart. On your report sheet record the plusses and the percentage of glucose indicated on the color chart. Repeat the procedure with the remaining samples. Calculate and record the percentages of glucose in samples B, C, and D assuming no glucose present in the original urine specimen. Even if the original urine is sugar-free, your observed and calculated results for glucose may not agree. (Why?)

NOTE: The mixture in the test tube must be closely observed during the reaction and the 15-second waiting period to detect the rapid *"pass-through"* color caused by glucose in excess of 2%. Should the color rapidly "pass-through" green, tan, and orange to a dark greenish-brown, record as over 2% glucose without comparing the final color in the tube with the color chart.

When glucose is *absent* (negative test) the solution in the test tube remains blue at the end of the waiting period. A *white* sediment which may form is to be ignored.

When glucose is *present* (positive test) the color changes will be greater with increasing amounts of sugar and the changes occur more rapidly. Color changes occurring after the 15-second waiting period should be disregarded. Bear in mind that a false positive test can occur in the presence of the interferences discussed in the introduction, such as vitamin C.

B. ***Detection of Glucose in Urine with CLINISTIX or TES-TAPE.*** Avoid touching the test areas of CLINISTIX or TES-TAPE that come in contact with the urine specimen. Prevent contamination from the work area or from volatile fumes. To retain the sensitivity of the test papers, protect them from moisture, sunlight, and heat. Freshly-voided urine samples are recommended. If the testing for glucose is delayed, the specimen should be refrigerated, but it should be brought back to room temperature before making the test. (Why?) Acid should not be added to the urine. The following preservatives used in their recommended quantities do not interfere: toluene, formalin, thymol, and Urokeep. Any glassware used should be thoroughly clean and free of disinfectants, acids, bases, and detergents.

Using a small *clean* beaker, test the following urine samples one at a time:
A — the original urine sample
B — 9 drops urine + 3 drops 1.0% glucose
C — 6 drops urine + 6 drops 1.0% glucose
D — 6 drops urine + 4 drops 0.5% vitamin C + 2 drops 1.0% glucose

Proceed with sample A as follows:

1. Dip the test area of CLINISTIX, or the end of a 4-cm piece of TES-TAPE, briefly in the urine sample. Remove immediately to avoid dissolving out the reagents in the test paper.

2. Tap the edge of the strip to remove excess urine.

3. Ten seconds after wetting the CLINISTIX, compare the color of the test area with the color chart on the CLINISTIX bottle. When using TES-TAPE, wait one minute before making final color comparison. On your report sheet record the percentage of glucose indicated by the color chart on the CLINISTIX bottle or the TES-TAPE dispenser.

Repeat the sugar test on samples B, C, and D. Calculate and record the percentage of glucose in the fortified urine samples B, C, and D.

REFERENCE

FELDMAN, J. M., KELLEY, W. M., and LEBOVITZ, H. E. 1970. Inhibition of glucose oxidase paper tests by reducing metabolites. *Diabetes* 19:337.

Reagents and Materials are listed on page 277.

Experiment 13. DETECTION OF GLUCOSE IN URINE

Calculate and record the percentages of glucose in samples B, C, and D, which you prepared by adding the 1.0% glucose specified.

Record the actual percentages you found by comparing the samples with the color charts.

Sample	CLINITEST TABLETS Percentage Glucose Calculated	Found	CLINISTIX or TES-TAPE Percentage Glucose Calculated	Found
A	. . . .	_____	. . . .	_____
B	_____	_____	_____	_____
C	_____	_____	_____	_____
D	. . . .	_____	_____	_____

Calculations for Samples B, C, and D.

CLINITEST Tablets. The color chart used with the CLINITEST represents the percent of glucose in the *undiluted* urine and *not* in the diluted mixture of 5 drops of urine plus 10 drops of water. Therefore, the *calculated values* for samples B and C must represent the percent of glucose in 5 drops of urine with the assumption that the *added* glucose was present in the original urine specimen. One way to obtain the calculated values in samples B and C directly is as follows:

$$\% \text{ glucose in B or C} = \frac{\text{drops of 1.0\% glucose}}{5 \text{ drops}} \times 1.0\%$$

CLINISTIX Test. The color chart used with the CLINISTIX test represents the percent of glucose in original urine *without* addition of water. Therefore, the *calculated values* for samples B, C, or D in this test must be based on the diluted mixture as if they constituted the original urine. The calculated values are obtained as follows:

$$\% \text{ glucose in B, C, or D} = \frac{\text{drops of 1.0\% glucose}}{\text{total drops in mixture}} \times 1.0\%$$

Vitamin C Addition. How did the tests on sample D (containing vitamin C) appear as compared to the tests with each of the following samples?

Sample A in CLINITEST (no glucose present) _____

_____

Sample B in CLINISTIX (glucose added) _____

_____

Name_____ Section_____ Date_____

Experiment 13. DETECTION OF GLUCOSE IN URINE

1. Give the chemical function of each of the chemical constituents of the CLINITEST tablets.

   Cupric sulfate pentahydrate _____

   _____

   Sodium hydroxide _____

   _____

   Sodium carbonate _____

   _____

   Citric acid _____

   _____

2. Using structural organic formulas where applicable, write *balanced* equations for the following reactions:
   a. Between the caustic soda and the acidic component.

   b. Between the constituents that produce the effervescence during the CLINITEST reaction.

3. Normally Cu^{++} ions in an alkaline solution containing carbonate ions will form a precipitate of basic cupric carbonate in which some of the CO_3^{2-} ions of the basic $CuCO_3$ have been replaced by OH^- ions. How is this precipitation of Cu^{++} ions prevented during the CLINITEST reaction? (Refer to Expt. 10.)

   _____

   _____

4. a. Give the complete formula of the component responsible for the *spotted blue color* of the CLINITEST tablets. _____

   b. Discuss the cause and the usefulness of tablets that have acquired an *overall blue* color.

   _____

   _____

5. Give the formula of the chemical substance in the CLINITEST reaction with urine responsible for

   a. The blue color of a negative test. _____

   b. The red-orange color of a positive test. _____

6. Both the CLINITEST and CLINISTIX methods involve an oxidation-reduction type of reaction.

   a. Name the oxidation product of glucose. _____

   b. Give the chemical formula of the *oxidizing agent* in the CLINITEST reaction _____ ;

   CLINISTIX reaction _____ .

   c. If glucose is the reducing agent in both test methods, what is the chemical formula of the reduction product formed in the CLINITEST reaction _____ ; CLINISTIX reaction? _____

7. Although both test methods involve the same type of reaction (oxidation-reduction), what is the fundamental difference between the CLINISTIX/TES-TAPE reactions and the CLINITEST reaction?

   _____

8. Compare the *specificity* of the two test methods toward reducing sugars.

   _____

   _____

9. Why would it be improper to use a *cold*, refrigerated urine specimen for detection of glucose by the CLINISTIX or TES-TAPE method without first bringing the sample to room temperature?

   _____

10. Explain briefly why vitamin C or the metabolites from certain drugs are likely to give false test results when they are present in a urine specimen.

    a. A false *positive* test with CLINITEST tablets in *absence* of glucose.

    _____

    _____

    b. A false *negative* test with CLINISTIX in *presence* of glucose.

    _____

    _____

Determination of Glucose in Blood Plasma

INTRODUCTION

Blood plasma or serum is generally used for the determination of glucose. Plasma is the clear fluid obtained when freshly-drawn blood is treated with an anticoagulant to prevent clotting and is then centrifuged to remove suspended constituents. Potassium oxalate is an anticoagulant that precipitates the calcium ions essential for normal clotting, and the added sodium fluoride acts as a preservative by preventing the glycolysis of glucose. In contrast, blood serum is obtained when whole blood is permitted to clot and the fibrin and the cellular components are removed by centrifugation.

In the past the Somogyi-Nelson method for the determination of glucose in a protein-free filtrate from blood plasma or serum was widely used. More recently, rapid colorimetric procedures using o-toluidine or enzymes, such as glucose oxidase and peroxidase, have replaced the Somogyi-Nelson method primarily because glucose can be determined directly on the plasma or serum. In these current methods the preparation of a protein-free filtrate is unnecessary unless the sample is turbid, icteric, or hemolyzed.

The addition of zinc sulfate and barium hydroxide to a diluted blood plasma or serum results in the precipitation of zinc hydroxide and barium sulfate. This precipitate absorbs the blood proteins and other reducing substances except glucose.

$$ZnSO_4 + Ba(OH)_2 \rightarrow \underline{Zn(OH)_2} + \underline{BaSO_4}$$

The concentrations of the two reagents have been previously adjusted so that no excess of either reagent is present in the filtrate. The protein-free filtrate contains few interferences and does not undergo a pH change.

In the Somogyi-Nelson procedure for glucose, a portion of the protein-free filtrate is heated with the alkaline copper reagent causing the glucose to be oxidized and the cupric ion to be reduced to insoluble cuprous oxide. The addition of the arsenomolybdate reagent dissolves and reoxidizes the Cu_2O, while the molybdenum in the arsenomolybdate is reduced to give a complex with a greenish-blue color. The intensity of the color of the solution is proportional to the amount of glucose present in the Folin-Wu tube.

The reagent used in the *glucose-oxidase method* is a buffered mixture of two enzymes (glucose oxidase and peroxidase) and a chromogen (o-dianisidine). When a portion of blood plasma or serum reacts with the test reagent, the following reactions occur:

$$\text{D-glucose} + H_2O + O_2 \xrightarrow{\text{glucose oxidase}} \text{D-gluconic acid} + H_2O_2$$

$$H_2O_2 \xrightarrow{\text{peroxidase}} H_2O + [O] \text{ active oxygen}$$

$$\begin{array}{c} \text{o-dianisidine (reduced)} + [O] \\ \text{(colorless)} \end{array} \longrightarrow \begin{array}{c} \text{o-dianisidine (oxidized)} \\ \text{(brown)} \end{array}$$

The intensity of the brown color of the test solution is proportional to the amount of glucose in the original sample.

The glucose-oxidase procedure is supposedly specific for glucose. However, investigators have found that the enzyme also catalyzes the oxidation of 2-deoxy-D-glucose, D-mannose, and D-fructose, but at a much slower rate. A number of interfering substances in the copper-reduction method and the glucose-oxidase test for glucose in urine have been discussed in the previous experiment. (Review the introduction to Expt. 13.) Some substances, such as uric acid and ascorbic acid, do not seriously interfere unless they are present in abnormally high levels. Deproteinization of blood plasma or serum must be done in the Somogyi-Nelson method because proteins interfere in the copper-reduction reaction, and it must also be done to clarify a turbid or colored plasma or serum sample used in the glucose-oxidase procedure. Fortunately, deproteinization yields a protein-free filtrate from which many other interfering substances have been also eliminated.

PROCEDURE

Check with your instructor to find out which part of this experiment you will perform. For example, you will prepare a protein-free filtrate of blood plasma (part A) if you do the Somogyi-Nelson method for glucose (part B). You do *not* need to prepare a protein-free filtrate (part A) if you are to use the *direct* glucose-oxidase procedure (part C).

Your instructor will provide you with an unknown blood plasma or serum sample. Alternatively, if desired, and proper arrangements can be made, the student may use his or her own blood. In this case 2 mg $K_2C_2O_4$ and 2.5 mg NaF (see Reagents p. 278) are dissolved in each ml of drawn blood, which is then centrifuged to obtain the blood plasma.

A. *Preparation of Protein-Free Filtrate.* Carefully follow each step of the procedure outlined below. In *step 1*, you will either be provided by your instructor with exactly 0.50 ml of an unknown sample, or you will pipet 0.50 ml of your own blood plasma.

1. A 0.50-ml sample of blood plasma is pipetted into a *clean, dry* 50-ml Erlenmeyer flask.
2. *Add exactly* 5.50 ml distilled water from the buret provided. *Mix* thoroughly.
3. Add *exactly* 2.0 ml barium hydroxide from the 25-ml buret provided. *Mix* well.
4. Add *exactly* 2.0 ml zinc sulfate from the 25-ml buret provided. *Mix* well. Let stand at least 5 min.
5. Filter the plasma mixture through a dry Whatman No. 40 filter paper placed in a *dry* long-stem funnel, whose tip is inserted into a clean, *dry* test tube. The funnel and test tube can be dried with a few ml of acetone, followed by a stream of air. Use your wooden funnel support attached to a ring stand for filtration process. Alternatively, the sample can be centrifuged.

The protein-free filtrate collected in the test tube must be perfectly *clear*. If not, check with your instructor. If the analysis cannot be completed, stopper the test tube and refrigerate the filtrate until the next lab period.

B. *Determination of Blood Glucose by the Somogyi-Nelson Method.* Pipet into each of seven Folin-Wu tubes (appropriately labeled) the volumes of solution specified in the following steps:

1. Pipet 1 ml of the clear protein-free filtrate into each of two Folin-Wu tubes U1 and U2. (The 1-ml transfer pipet can be dried with acetone and air.)
2. In a clean, *dry* test tube obtain 3-5 ml of dilute standard glucose (0.20 mg glucose/ml). Using a 1-ml graduated pipet, add to five Folin-Wu tubes the quantities of water indicated in the table below. Dry the same 1-ml pipet with acetone and air, and add the volumes of standard glucose shown.

Tube No.	Blank	1	2	3	4
ml distilled water	1.00	0.80	0.70	0.50	0.30
ml standard glucose	....	0.20	0.30	0.50	0.70

3. Add 1.0 ml of a fresh solution of copper reagent (p. 278) to each of the seven Folin-Wu tubes using the 10-ml buret provided.

4. Place all the tubes in an *actively* boiling water bath for 20 minutes.

5. Cool in beaker of cold tap water.

6. Add 1.0 ml of arsenomolybdate from the 10-ml buret provided to each of the Folin-Wu tubes. *Mix well.*

7. Add distilled water exactly to the 25-ml mark on the tubes. *Mix thoroughly by 6-8 complete inversions of the tubes.*

8. Read the color intensities of the glucose standards and the unknown in the Spectronic 20 colorimeter using a wavelength of 520 nanometers. Set the instrument at *zero* Absorbance (100% Transmittance) by using the "Blank" solution in a cuvette. Your instructor will teach you the proper use of the instrument. Read "Notes on the Use of the Spectronic 20" on page 21, and review "Colorimetry" in the introduction to Experiment 4.

9. Record absorbancies on the report sheet and plot the standard curve.

C. **Direct Determination of Blood Glucose by the Glucose-Oxidase Method.** Your instructor will provide you with exactly 0.50 ml of an unknown sample in a test tube, or you will pipet 0.50 ml of your own blood plasma in a clean, *dry* test tube.

Prepare a *diluted* sample by adding *exactly* 9.50 ml of distilled water to the 0.50-ml sample in the test tube. Cover with Parafilm and mix thoroughly. (**NOTE:** Do *not* use rubber stoppers to seal tubes or reagents because certain substances in rubber may inhibit enzyme reactions.)

Label five clean, *dry* test tubes as follows: B (blank), S-1, and S-2 (standards), and U-1 and U-2 (unknowns). Carefully follow the procedure outlined below, measuring all volumes accurately.

1. Pipet 0.50 ml of water into the Blank tube, and 0.20 ml of water into tube S-1 only, using the 1-ml graduated pipet.

2. *Dry* the graduated 1-ml pipet with acetone and air. Obtain 3-5 ml of standard glucose solution (0.10 mg/ml) in a clean, *dry* test tube.

3. Using the dry 1-ml graduated pipet, add 0.30 ml of standard glucose solution to tube S-1 and add 0.50 ml to tube S-2.

4. Thoroughly rinse and dry the 1-ml graduated pipet. Pipet exactly 0.50 ml of the *diluted* unknown sample (prepared earlier) into each of the unknown tubes, U-1 and U-2.

5. Add exactly 5.00 ml of the combined enzyme/color reagent (available in a buret on the reagent bench) to each of the five tubes. Cover the tubes with Parafilm and mix the contents of each tube thoroughly by repeated inversions.

6. Allow the tubes to stand for 45 minutes at room temperature in your desk cabinet, or incubate at 37° for 30 minutes. *Avoid exposure to direct sunlight or bright daylight.*

7. Read the absorbancies (A) of the glucose standards and the unknowns with the Spectronic 20 colorimeter, using a wavelength of 450 nm. Set the instrument at *zero* Absorbance (100% Transmittance) by using the blank solution in a cuvette. Your instructor will teach you the proper use of the instrument. Readings should be completed within 30 minutes. Read "Notes on the Use of the Spectronic 20" on page 21 and review "Colorimetry" in the introduction to Experiment 4.

8. Record your data on the **report sheet** and calculate your unknown glucose values.

Calculations. It has been found that a plot of absorbance versus glucose concentration is a *straight line* up to a glucose level of 300 mg/100 ml plasma, indicating that Beer's law is obeyed. Consequently, the ratio of the concentrations of the unknown and standard is directly proportional to their respective absorbancies.

$$\frac{C_u}{C_s} = \frac{A_u}{A_s} \quad \text{or,} \quad C_u = \frac{A_u}{A_s} \times C_s$$

where C_u = glucose conc. of unknown in mg/100 ml
C_s = glucose conc. of standard in mg/100 ml
A_u = absorbance of unknown
A_s = absorbance of standard

From the above equation, the glucose concentration can be calculated in terms of mg/100 ml plasma or serum. Calculate the amount of glucose in your unknown on the basis of both standards, S-1 and S-2, as a check on your analytical technique.

REFERENCES

FALES, F. W. 1963. *Standard methods of clinical chemistry,* ed. David Seligson, vol. 4, pp. 101-12. New York: Academic Press.
NELSON, M. 1944. *J. Biol. Chem.* 153: 375.
RAABO, E., and TERKILDSEN, T. C. 1960, *Scan. J. Clin. Lab. Invest.* 12:402.
SOMOGYI, M. 1945. *J. Biol. Chem.* 160: 69.

Reagents, Materials, and Equipment are listed on page 277.

Name _____ Section _____ Date _____

Experiment 14.   DETERMINATION OF GLUCOSE IN BLOOD PLASMA

Unknown Sample No. _____

B. *Determination of Blood Glucose by the Somogyi-Nelson Method.*

In the preparation of the protein-free filtrate, 0.50 ml of blood plasma is diluted with water and reagents to a final volume of 10 ml.

a. If 1 ml of filtrate is used, how many ml of *original* blood plasma is present in the Folin-Wu tube? Show work.

_____ ml blood plasma/tube

b. By what factor must the mg of glucose found in the Folin-Wu tubes (unknown and standards) be multiplied to express the results as mg glucose per 100 ml blood plasma? Show work.

The dilution factor is_____.

Standard Curve

The standard curve is prepared by plotting the absorbancies of the standards on the vertical axis (ordinate) against the glucose concentrations of the standards on the horizontal axis (abscissa). The horizontal axis can be marked off as a double scale; one scale is expressed as mg glucose per standard tube, and the other scale below it as mg glucose per 100 ml of blood plasma (mg glucose/tube multiplied by the dilution factor). Thus, one can merely take the absorbance reading of the unknown and immediately read from the curve the mg glucose per 100 ml blood plasma. Attach your graph to the lab report.

Tube No.	Glucose standards				Unknown	
	1	2	3	4	U-1	U-2
Absorbance						
Ml standard glucose*					-----	-----
Mg glucose/standard tube					-----	-----
Mg glucose/100 ml plasma**						

*Standard glucose = 0.20 mg/ml.
**Mg glucose/standard tube × dilution factor.

C. *Direct Determination of Blood Glucose by the Glucose-Oxidase Method*

In the preparation of the diluted plasma sample, 0.5 ml of blood plasma is mixed with 9.5 ml of water to a final volume of 10 ml.

a. If 0.5 ml of the diluted sample is used, how many ml of the *original* blood plasma is in tube U-1 and U-2? Show work.

_____ ml blood plasma/tube

b. By what factor must the mg of glucose found in the test tubes (unknown and standards) be multiplied to express the results as mg glucose per 100 ml blood plasma? Show work.

The dilution factor is_____

	Tube S-1	Tube S-2	
Ml standard glucose	_____	_____	(Standard glucose: 0.1 mg/ml)
Mg glucose/standard tube	_____	_____	
Mg glucose/100 ml plasma	_____	_____	(Mg glucose/tube × dilution factor)

Absorbance   S-1 _____   S-2 _____   U-1 _____   U-2 _____

Calculations for Blood Plasma Glucose (mg/100 ml)

Calculate the glucose in the original unknown sample using your data and the equation:

$$C_u = \frac{A_u}{A_s} \times C_s$$

Use *both* standards, S-1 and S-2, for calculating the glucose value from absorbance of U-1 and average the two values. Do the same using the absorbance for U-2.

	Unknown U-1		Unknown U-2	
	S-1	S-2	S-1	S-2
Based on standard				
Glucose, mg/100 ml plasma	____	____	____	____
Average value	____		____	

Name _____ Section _____ Date _____

Experiment 14. DETERMINATION OF GLUCOSE IN BLOOD PLASMA

1. In the preparation of blood plasma from freshly-drawn blood,
   a. How is the blood prevented from clotting?

   _____

   b. What is the function of NaF?

   _____

2. Why is it necessary to use *dry* glassware at certain points in preparing the glucose standards or in the preparation of a protein-free filtrate (when required)?

   _____

3. Why are proteins removed in the Somogyi-Nelson method? Be specific.

   _____

4. Under what circumstances *must* a protein-free filtrate be prepared in the glucose-oxidase method?

   _____

5. Name the two *principle* blood proteins removed by the $Zn(OH)_2$ and $BaSO_4$ precipitate.

   _____ and _____ .

6. Briefly state how you can determine if the $Ba(OH)_2$ and $ZnSO_4$ solutions are properly balanced (react stoichiometrically)? (See Reagents in Appendix.)

   _____

   _____

   _____

   _____

7. In both the Somogyi-Nelson and the glucose-oxidase methods, glucose is oxidized to gluconic acid. What is the *oxidizing agent* and *reduction product*

   a. In the Somogyi-Nelson method?

   _____

   b. In the glucose-oxidase method?

   _____

8. What is the purpose of setting up a reagent blank together with the unknown and glucose standards?

   _____

   _____

9. What are the advantages of the *direct* glucose-oxidase method over the Somogyi-Nelson method?

_____

10. What is the normal range of glucose in a "fasting" blood plasma sample?

_____ mg/100 ml

11. Glucose is determined in urine (Expt. 13) and in blood every half hour for about 3 hours during a "glucose tolerance test." Briefly state the purpose of a "glucose tolerance test" and what glucose levels are expected in urine and blood during the test period. (Consult a textbook.)

Quantitative Determination of Sugars in Foods

INTRODUCTION

The sugars primarily present in foods are glucose, fructose, sucrose, and lactose. Glucose and fructose are commonly found in fruits and fruit juices, in plant saps, in honey, and in equal amounts in "invert sugar," which is made by the hydrolysis of sucrose. Generally, sucrose occurs naturally in low amounts in most fruits and vegetables, but is the main carbohydrate in sugar cane and sugar beets from which it is extracted and purified to yield our "table sugar." Frequently, it is added to foods as a sweetener. Since sucrose is a nonreducing disaccharide, it is necessary in the course of a sugar analysis to convert it into its components, glucose and fructose, by hydrolysis with dilute sulfuric acid. Lactose, a reducing disaccharide, is present in milk and milk products, and since it is the only reducing sugar generally present, except in sweetened condensed milk, it is determined directly as lactose without the use of acid hydrolysis in the sample preparation.

Of the many proposed methods for the detection and quantitative determination of sugars in foods, the procedure using an alkaline solution of cupric ions, which originated over 100 years ago, is still one of the methods commonly used today with some modifications and variations. In recent years, however, more sophisticated and expensive instrumental techniques have been developed, such as gas-liquid chromatography (GLC) and high pressure liquid column chromatography (HPLC). By using HPLC, for example, a complete quantitative analysis for each mono-and disaccharide in an aqueous extract of a mixture of sugars can be done in about 15 minutes.

In this experiment a suitable extract of the sample is allowed to react with a copper(II) solution, which will oxidize the sugars, and a portion of the cupric ions will be reduced to cuprous oxide (reducing sugar $+ 2Cu^{++} \rightarrow Cu_2O +$ oxidized sugar acids). When the solution is acidified with dilute sulfuric acid, the cuprous oxide dissolves ($Cu_2O + 2H^+ \rightarrow 2Cu^+ + H_2O$), and iodine is liberated from a precise amount of potassium iodate and potassium iodide already present in the copper reagent ($ID_3^- + 5I^- + 6H^+ \rightarrow 3I_2 + 3H_2O$). The cuprous ions are then reoxidized by the liberated iodine ($2Cu^+ + I_2 \rightarrow 2Cu^{++} + 2I^-$) and the excess iodine is measured by titration with standard sodium thiosulfate solution ($I_2 + 2S_2O_3^= \rightarrow 2I^- + S_4O_6^=$). A blank titration is done in exactly the same manner, except that no sugar sample is present. The difference between the two titrations gives the amount of iodine used to oxidize the Cu^+, and hence is a measure of reducing sugar that was present in the sample.

PROCEDURE

A. *Preparation of Sample.* (Consult the bulletin board for your assigned sample. Your sample should be run in duplicate).

(1) *Milk and milk products.* The amount of sample taken should be approximately as follows: whole milk or skim milk, 2.0 g; evaporated milk, 1.0 g; whole milk powder or skim milk powder, 0.3 g. These quantities will need to be weighed accurately on the quantitative balance. Transfer the sample without any loss into a 250-ml volumetric flask, rinsing the liquid samples into the flask completely. Add enough water so that the total volume used will be about 50 ml and mix thoroughly.

To remove milk proteins that interfere with the determination of sugar in part B, add 5 ml of 5% $CuSO_4$ and 10 ml of 0.1 M NaOH and mix well. These volumes do not need to be accurately measured, so use a graduated cylinder. Then fill the flask to the mark with water, mix well, filter a 75-100-ml portion through a dry paper and funnel into a dry 125-ml Erlenmeyer flask, and save the filtrate for sugar determination as indicated under B. Note that the sugar being determined in milk and milk products is lactose and that this requires a somewhat different handling under B and C.

(2) *Honey and syrups.* Make a small cup by shaping a one-inch-square piece of aluminum foil over the small end of a number 2 cork and pressing against a flat surface so that the cup will stand upright. Weigh the cup accurately (to the nearest 0.001 g) on the quantitative balance and then weigh the sample of honey or syrup in it. Be absolutely certain that none of the sample escapes from the cup. The quantity of sample taken should be in the range of 0.2 g. Drop the cup containing the sample into a clean 125-ml Erlenmeyer flask, add exactly 50 ml of water (pipet), and swirl or stir with a stirring rod until you are certain that the sample is completely dissolved and uniformly mixed into the solution. Then pipet a 5-ml aliquot into a 50-ml volumetric flask, add about 5 ml of 0.5 M H_2SO_4 (measured in a graduated cylinder) and heat the flask 10 minutes in a boiling water bath. Cool under the water tap, add a few drops of phenolphthalein, and neutralize with 1 M NaOH (faint pink color). Dilute to the mark with water, mix thoroughly, and use an aliquot for sugar determination as directed under B.

(3) *Fruit and vegetable juices.* Pipet 5 ml (5 g) of the juice directly into a 250-ml volumetric flask, dilute with water to the mark. Stopper the flask and mix thoroughly by repeated inversions and shaking. Pipet a 5-ml aliquot into a 50-ml volumetric flask, add about 5 ml of 0.5 M H_2SO_4, and proceed as described under (2) above, line 8.

(4) *Solid fruits and vegetables.* Cut the solid raw fruit or vegetable into small pieces and weigh a quantity in the range of 5 g for a fruit or 10 g for a vegetable on a weighed piece of aluminum foil to the nearest 0.01 g and transfer to a 250-ml beaker. Add about 25-30 ml of water, boil for 10 minutes, and strain through a piece of clean cheesecloth in a funnel into a 250-ml volumetric flask. To the insoluble residue add a second 25 ml of water, boil again for 2-3 minutes, and strain the liquid into the volumetric flask. Repeat a third time with about 50 ml of water. Cool the flask under the water tap and then dilute to the mark with water.

Mix the solution in the stoppered volumetric flask very thoroughly by inverting and shaking repeatedly. Then pipet a 10-ml aliquot into a 50-ml volumetric flask, add about 5 ml of 0.5 M H_2SO_4 and proceed as described under (2) above, line 8.

B. **Determination of Reducing Sugar.** A 5-ml aliquot of the sample solution prepared as described previously is *pipetted* into a 25 x 200-mm test tube and exactly 5 ml of the copper reagent added, using the buret provided on the reagent bench. The contents of the tube are well mixed, the tube is *loosely* stoppered, immersed in a boiling water bath and heated for *exactly* 15 minutes. Have the water boiling before the tube is put into the bath. (**NOTE:** In the case of the milk samples in which the sugar being determined is lactose, the heating period in the presence of the copper reagent must be extended to 25 minutes, since lactose reacts more slowly than other sugars with the copper reagent.) Strictly adhere to the specified heating time because the reaction between the alkaline Cu^{++} ions and the reducing sugars is not stoichiometric.

The tube is then removed, cooled under the water tap, and about 5 ml of 0.5 M H_2SO_4 (measured in a graduated cylinder) is added. The tube is allowed to stand for 1-2 minutes with occasional swirling and the liberated iodine then titrated with 0.005 N sodium thiosulfate solution until the yellow-brown iodine color is almost gone. A few drops of starch indicator are then added and the titration continued until the blue starch iodine color is just discharged. Note, however, that the solution will still have a very faint blue color because of the presence of unused copper reagent. Record the volume of thiosulfate solution used. The entire procedure is repeated on a second 5-ml aliquot of the prepared sugar samples. The two titrations with 0.005 N $Na_2S_2O_3$ should check reasonably well.

A blank determination must be carried out in exactly the same way, except that 5 ml of water are used in place of the 5 ml of sample solution. The titration in this case should require about 22 ml of the 0.005 N thiosulfate. (The instructor may choose to furnish you with the blank titration value.)

The difference between this blank titration figure and the titration of the sample solution represents the iodine utilized in reoxidation of the copper, which was originally reduced by the sugar. This difference in the two titrations (i.e., ml of 0.005 N thiosulfate) is converted into the corresponding weight of reducing sugar by reference to a calibration curve, which will be supplied by the instructor. This curve is made by carrying out the determination with known amounts of pure glucose or lactose.

C. **Calculation and Recording of Results.** On the laboratory report sheet first record the name of the sample that you have analyzed. Tabulate neatly all the data involved in determining the weight or volume of the sample taken. Summarize the different dilutions made and aliquots removed at the various stages of the analysis. Record the volume in ml of thiosulfate solution required for titration of the blank and that required for the sample. Take the difference in the two titration values and record the amount of glucose or lactose corresponding to this volume as read from the proper standard curve. Calculate the weight of sugar in the original weight of sample taken for analysis, taking into consideration the dilutions made and aliquots used at various stages of the analysis. Finally calculate the percentage of sugar in the sample. If you require help for this calculation, consult the laboratory instructor.

D. **Calibration Curves.** The calibration curves for lactose and glucose will be prepared by the instructor and posted on the bulletin board. The assistance of students with competence in analytical technique might be solicited. It is recommended that each sugar be run in duplicate.

(1) *Lactose.* Into 25 x 200-mm test tubes pipet 0, 1, 2, 3, 4, and 5 ml of a lactose solution containing 0.75 mg/ml lactose (see Reagents, p. 280). To the respective test tubes pipet 5, 4, 3, 2, 1, and 0 ml of distilled water so that each tube now contains 5 ml of solution. Add exactly 5 ml of the copper reagent to each tube, and heat the tubes for *exactly 25* minutes in a boiling water bath. Cool, add the 5 ml of 0.5 M H_2SO_4 and proceed with the titration of iodine as described in part B. The test tube without lactose represents the *blank titration*. The volume of 0.005 N thiosulfate required for each sugar-containing tube is subtracted from the blank titration. These titration differences (B—S) are plotted on the vertical axis of a graph paper against the weights of lactose in the tubes (0, 0.75, 1.50, 2.25, 3.00, 3.75 mg lactose, respectivly).

(2) *Glucose.* Using a glucose solution containing 0.45 mg/ml glucose (see Reagents, p. 280) set up the test tubes as described for lactose in (1) above. Note, however, that the heating time with the copper reagent is *exactly 15* minutes. Plot a graph with the titration differences (B—S) against the weights of glucose in each tube, which in this case are 0, 0.45, 0.90, 1.35, 1.80, and 2.25 mg glucose, respectively.

REFERENCES

POMERANZ, Y., and MELOAN, C. E. 1978. *Food analysis: theory and practice.* Rev. ed., chapter 35. Westport, Conn.: AVI Publishing Co., Inc.

Reagents and Materials are listed on p. 279.

LABORATORY REPORT

Name _____ Section _____ Date _____

Experiment 15. QUANTITATIVE DETERMINATION OF SUGARS IN FOODS

Name of sample analyzed: _____

Sample data:

Fruit juice: Volume = 5.00 ml; weight = 5.0 grams

	1.	2.
All other samples:		
Weight of sample + container	_____	_____
Weight of container	_____	_____
Weight of sample, grams	_____	_____

Dilution and aliquot data:

Original volume to which sample was diluted _____ ml

Aliquots taken at various stages:

Titration data: (Volumes of 0.005 N sodium thiosulfate)

	1.	2.	
Blank determination (B)	_____	_____	ml
Excess iodine titrated (S)	_____	_____	ml
M1 $Na_2S_2O_3$ consumed by sample (B—S)	_____	_____	ml

Percentage reducing sugar:

Grams (specify sugar_____) from standard curve for aliquot analyzed .... _____ _____ grams

Calculations of percentage sugar in original sample: (show work)

_____ %	_____ %

QUESTIONS

Name_____ Section _____ Date _____

Experiment 15.   QUANTITATIVE DETERMINATION OF SUGARS IN FOODS

1. Why are solutions of 5% $CuSO_4$ and 0.1 M NaOH added in the preparation of the milk or milk product solutions but not in the sample preparation for honey, syrups, fruits, and vegetables?

_____

_____

2. a. Why is it necessary to heat the sample solution with 0.5 M H_2SO_4 in the case of honey, syrups, fruits, and vegetables before the actual sugar analysis in part B is carried out?

_____

   b. What sugars would be measured if this step were omitted?

_____

3. How can the amount of sucrose in a sample of food be determined by a modification of the sugar analysis procedure described in this experiment? Give principle of method only.

4. Write the balanced, net ionic equation for the involvement of dilute H_2SO_4 in the following reactions:

   a. Dissolving cuprous oxide._____

   b. Liberation of iodine. _____

5. Write balanced, net ionic equations for the reactions of iodine with the following:

   a. With cuprous ion. _____

   b. With sodium thiosulfate._____

6. For preparing the lactose calibration curve, 1.5 g of lactose are dissolved in water and made to 100 ml. A 5-ml aliquot is then further diluted to 100 ml. Calculate the mg of lactose present in 1 ml and 5 ml of the second diluted solution. Show work.

_____ mg lactose/1 ml

_____ mg lactose/5 ml

Lipids

Esters

INTRODUCTION

Esters are substances that can be formed by the interaction of alcohols or phenols with organic or inorganic acids. The fragrance of flowers and fruits is due primarily to the presence of esters. The component parts of many biological substances are joined together by ester linkages as, for example, in triacylglycerols (triglycerides), phospholipids, sugar phosphates, and the nucleic acids. The student, therefore, should be thoroughly familiar with this chemical group, both in its formation and in its hydrolysis.

The reaction between an alcohol and an organic acid seldom goes to completion, since the formation of water is such that the rates of esterification and the hydrolysis of the ester reach an equilibrium.

$$\underset{\text{O}}{\overset{\text{O}}{\|}}$$

$$\text{R-C-OH} + \text{HO-R'} \underset{\text{hydrolysis}}{\overset{\overset{\text{H}^+/\text{heat}}{\text{esterification}}}{\rightleftharpoons}} \text{R-C-O-R'} + \text{H}_2\text{O}$$

where R represents an alkyl or aromatic group of an organic acid and R'-OH is an alcohol or phenol. The reaction is catalyzed by the H^+ ion from a small amount of concentrated sulfuric acid added to the reaction mixture. The preparation of *ethyl acetate* illustrates the synthesis of one of the many simple organic esters.

Salicylic acid has both the carboxyl and the phenolic groups attached to the benzene ring. In the preparation of *methyl salicylate* (oil of wintergreen), the carboxyl group of salicylic acid interacts with methyl alcohol. In the synthesis of *acetylsalicylic acid* (aspirin), however, the phenolic group of salicylic acid combines with an acetyl group from acetic anhydride. The latter acetylating reagent reacts more rapidly with the phenolic group than acetic acid does directly. Furthermore, no water is formed to cause hydrolysis of the aspirin; rather, acetic acid is a by-product so that good yields of aspirin are obtained.

salicylic acid + acetic anhydride $\xrightarrow[\text{heat}]{\text{H}^+}$ acetylsalicylic acid (aspirin) + CH_3COOH acetic acid

The alkaline hydrolysis of an ester goes to completion due to the formation of the Na^+ or K^+ salt of the organic (or inorganic) acid.

$$\underset{\text{O}}{\overset{\text{O}}{\|}}$$

$$\text{R-C-O-R'} + \text{NaOH (or KOH)} \longrightarrow \text{R-C-O}^-\text{Na}^+ \text{ (or K}^+) + \text{R'-OH}$$

When the alkaline hydrolysis of a triacylglycerol, or other simple ester, is involved, the process is called *saponification*, which will be more closely examined in Experiment 17. In this experiment the alkaline hydrolysis of ethyl butyrate demonstrates the loss of its fragrance, which is ultimately replaced by the disagreeable odor of the short-chain organic acid liberated from its salt upon acidification of the alkaline mixture.

Of great biological importance are the phosphate esters resulting from the enzymatically-catalyzed reactions between compounds containing alcoholic —OH groups and the —OH group of phosphoric acid derivatives. The formation of phosphate mono- or diester compounds can be illustrated in a simple way with phosphoric acid, H_3PO_4, as follows:

$$R-OH + HO-\overset{\overset{\displaystyle O}{\|}}{\underset{\underset{\displaystyle OH}{|}}{P}}-OH \leftrightharpoons R-O-\overset{\overset{\displaystyle O}{\|}}{\underset{\underset{\displaystyle OH}{|}}{P}}-OH + H_2O \overset{ROH}{\rightleftharpoons} R-O-\overset{\overset{\displaystyle O}{\|}}{\underset{\underset{\displaystyle OH}{|}}{P}}-O-R + H_2O$$

phosphate monoester $\qquad$ phosphate diester

where R-OH represents a biological hydroxy compound, such as a monosaccharide, glycerol, choline, or ethanolamine. The resulting products may be metabolically important sugar phosphates (e.g., glucose-6-phosphate). Examples of substances with a phosphate diester grouping are the phospholipids, such as phosphatidyl choline (lecithin) or the nucleic acids, RNA and DNA, whose structural backbone consists of a chain of alternating pentose and phosphate units joined by 3′,5′phosphate diester bonds. The hydrolysis and identification of the components of nucleic acids will be taken up in Experiment 39.

PROCEDURE

In the esterification reactions given below, it is essential that the temperature of the reaction mixture does *not* exceed that specified, otherwise the alcohol (and perhaps also the ester) will volatilize and seriously reduce the yield of the ester. Start each reaction with a *dry* test tube to minimize hydrolysis of products. Show your instructor your results in each part.

A. *Ethyl Acetate.* Place in a dry test tube 2 ml each of *glacial* acetic acid and ethyl alcohol. Note the odor of each reagent when you do this. Add 3 to 5 drops of *concentrated* H_2SO_4 and warm the tube for ten minutes in a water bath at *70-80°C*. Cool, add 10 ml of cold tap water and about 3 grams of NaCl to "salt-out" the ethyl acetate. Stir to dissolve the NaCl, note the odor from the organic layer of the ester. Record your findings on the laboratory report sheet.

B. *Methyl Salicylate (Oil of Wintergreen).* Place 2 ml of methyl alcohol and 1 g of salicylic acid, $C_6H_4(OH)(COOH)$, in a dry test tube and add a few drops of concentrated H_2SO_4. Warm in a water bath at *50-60°C* for ten minutes. Pour the mixture into a watch glass and note the odor. The white residue on the watch glass is unreacted salicylic acid intermixed with oil of wintergreen. Note your results on the report sheet.

C. *Acetylsalicylic Acid (Aspirin).* Place 2 g of salicylic acid into a dry test tube and carefully add 4 ml of acetic anhydride, using the graduated cylinder provided in the reagent hood. (**CAUTION:** The reagent is a skin and nasal irritant.) Add 4 drops of concentrated H_2SO_4 to the mixture and stir thoroughly with a glass rod for several minutes *under your hood*. Warm the test tube in a hot water bath (70-80°C) until all of the solid material dissolves. Continue heating in the water bath at 70-80° with frequent stirring for 5 minutes to complete the reaction.

Set the tube aside to cool for several minutes. The aspirin should begin to crystallize and, if not, scratch the inside wall of the tube with the stirring rod. Then add 12-15 ml of cold water to hydrolyze the excess acetic anhydride, and set the tube in an ice-water bath (beaker) to complete the crystallization, stirring occasionally.

Place a *small* ball of glass wool in the apex of a glass funnel and moisten it with water. Suspend the aspirin by stirring the contents of the tube, and pour it on the glass wool to collect the aspirin. Rinse the test tube with small portions of cold, distilled water, pouring each successive rinsing into the funnel. Finally, rinse the funnel and the collected aspirin with small portions of cold water.

Transfer the aspirin to a 250-ml beaker, add 75 ml of distilled water, and slowly heat to 70-80°C, with constant stirring to dissolve the crystals. Prepare a wet filter paper in the glass funnel, and when most of the acetylsalicylic acid has dissolved, pour the hot solution into the filter, collecting the filtrate in a clean 150-ml beaker. Place the beaker in an ice-water bath for 15-20 minutes to recrystallize the aspirin. Filter the product through a clean, moist filter paper, wash with 10 ml of cold distilled water, in small portions, allowing each portion to drain from the funnel before adding the next. (**NOTE:** A Büchner funnel can be used if available.) Predry your product between pieces of filter paper or paper towel and complete the drying on a weighed watch glass. Record the weight of the dry aspirin.

Test for purity. Into separate test tubes add about 0.1 g of a crushed aspirin tablet, your prepared aspirin, and salicylic acid, respectively. Add 5 ml of water to each tube and shake to disperse as much as possible. Then add 2-3 drops of 0.1 M ferric chloride solution to each tube and shake. An orange to deep purple color is indicative of the presence of the phenolic group as in unreacted salicylic acid. Record your observations.

D. *Saponification of Ethyl Butyrate.* Place no more than 5-6 drops of ethyl butyrate in a test tube. Note and record the odor. Add 2 ml of 3 M KOH in 50% ethyl alcohol and heat in a water bath at 80-85°C for ten minutes with *frequent shaking.* Raise the temperature of the bath to 90-95° for 5-10 minutes to complete saponification. Cool to 40-50° and to the warm solution add 2 drops of 1% phenolphthalein indicator. Add 2 ml of 3 M HCl and mix. Continue adding 3 M HCl dropwise until the pink solution turns colorless. Note and record the odor of the liberated butyric acid. Write equations for the reactions on your report sheet. (To the instructor: A corked test tube containing a drop of butyric acid should be available for comparison.)

Reagents and Materials are listed on page 280.

LABORATORY REPORT

Name_____ Section_____ Date_____

Experiment 16.   ESTERS

A. *Ethyl Acetate*

1. Describe and compare the odors of the following:

   Acetic acid _____

   Ethyl alcohol _____

   Ethyl acetate _____

2. What evidences indicate that a reaction had occurred?

   _____

3. Using structural formulas write an equation for the esterification. Name each organic substance.

B. *Methyl Salicylate (Oil of Wintergreen)*

1. Describe the odor. _____
2. Write an equation for the preparation of methyl salicylate, including complete structure of salicylic acid and other organic substances.

C. *Acetylsalicylic Acid (Aspirin)*

1. Weight of watch glass and aspirin _____

   Weight of watch glass .... _____

   Weight of aspirin .... _____

2. Using structural formulas, write an equation for the preparation of aspirin. Name each substance.

3. Write an equation for the hydrolysis of excess acetic anhydride when water is added to the reaction mixture.

4. Describe the physical appearance of your aspirin.

_____

5. Describe the results of the ferric chloride test on the following:

   a. Crushed aspirin tablet _____

   b. Your prepared aspirin _____

   c. Salicylic acid _____

D. *Saponification of Ethyl Butyrate*
   1. Describe the odor of ethyl butyrate. _____

   2. Using structural formulas, write equations for
      a. The saponification of ethyl butyrate

      b. The acidification of the organic product from (a).

      c. Describe the odor _____ Compound causing odor _____

QUESTIONS

Name_____ Section_____ Date _____

Experiment 16.   ESTERS

1. What is the purpose of concentrated H_2SO_4 in esterification?

    _____

2. Using C and O atoms, illustrate what is meant by the ester group.

    _____

3. Give at least two commercial or pharmaceutical uses for each of the following esters:

    a. Ethyl acetate _____

    b. Methyl salicylate _____

4. Why is acetic anhydride used instead of acetic acid in the preparation of acetylsalicylic acid?

    _____

5. Give the names and structures of the specific *functional groups* of salicylic acid that are involved in the preparation of each of the following esters:

    a. Methyl salicylate _____

    b. Acetylsalicylic acid _____

6. Complete the following table by giving the common names of the esters and the names and structures of the alcohols and acids required to synthesize them.

Structure of Ester	Common Name	Reagents required
CH_3CH_2OCH with O double bonded		
$(CH_3)_2CHCH_2OCCH_2CH_3$ with O double bonded		
$CH_3(CH_2)_4OCCH_3$ with O double bonded		
$CH_3(CH_2)_2OC(CH_2)_2CH_3$ with O double bonded		

7. Write the structures of the following lipids:

    a. General structure of a triacylglycerol (triglyceride)

    b. Phosphatidic acid

    c. Phosphatidyl ethanolamine (as a dipolar ion)

8. Write the structures of the following phosphate esters: (Show complete phosphate structures.)

    a. α-D-glucose-6-phosphate (pyranose ring)

    b. Fructose-1, 6-diphosphate (furanose ring)

    c. The 3',5'-phosphate diester linkage between two ribose units as it occurs in a segment of the polynucleotide chain of RNA.

Saponification of Fats

INTRODUCTION

The chemical change involved in the saponification of fats is identical with the alkaline hydrolysis of ethyl butyrate (Expt. 16). However, in the present experiment, we are dealing with higher molecular weight fatty acids, saturated and unsaturated, and with a polyhydroxy alcohol, instead of a simple monohydroxy alcohol. The nature of the soap separated from the saponification mixture depends on the kind of alkali used and the type of fats saponified. Sodium hydroxide and solid fats tend to form hard soaps, while potassium hydroxide and fatty oils tend to yield soft soaps. Likewise the fatty acids liberated after the acidification of soap will be solid or semisolid, depending on the kind of fatty material saponified.

If the saponification of a fat is done quantitatively on a small scale (0.2-0.3 g) with an alcoholic solution of standard KOH, the *saponification number* obtained represents the number of milligrams of KOH required to saponify 1 gram of fat. This analytical value is not only useful to the soap manufacturer in determining his alkali requirements, but a high saponification number (e.g., 250 for coconut oil and 230 for butter) indicates to the chemist the presence of a large percentage of fatty acids containing 12 carbons or less, while a low value (e.g., 193 for cottonseed oil) is indicative of a large percentage of C_{16} and C_{18} fatty acids.

Naturally occurring fats and fatty oils are a mixture of mixed acylglycerols and, to a much lesser degree, simple acylglycerols. A mixed acylglycerol contains two or three fatty acids esterified to a glycerol unit. If three different fatty acids are present in a given molecule, it is called a *triacylglycerol* (formerly called a *triglyceride* until a recommended change by an international nomenclature committee). In a simple triacylglycerol, all three fatty acid acyl groups are the same. While the fatty acids may vary from butyric acid (C_4) to arachidic and arachidonic acids (C_{20}), the fatty acids that predominate are generally palmitic, stearic, oleic, and linoleic acids. Thus, fats and fatty oils are a very complex mixture of triacylglycerols, and the saponification reaction can be best illustrated by using a general formula to represent the triacylgercerol.

$$
\begin{array}{ccccccc}
& & \overset{\displaystyle O}{\overset{\displaystyle \|}{CH_2-O-C-R}} & & & & \\
\overset{\displaystyle O}{\overset{\displaystyle \|}{R'-C-O-C-H}} & & & + & 3NaOH & \rightarrow & \\
& & \overset{\displaystyle O}{\underset{\displaystyle \|}{CH_2-O-C-R''}} & & & &
\end{array}
$$

| Triacylglycerol (Triglyceride) | Glycerol | Na$^+$ salts of fatty acids (Soap) |

CH$_2$OH
|
CHOH
|
CH$_2$OH

RCOO$^-$Na$^+$
R'COO$^-$Na$^+$
R''COO$^-$Na$^+$

PROCEDURE

A. *Saponification of a Solid Fat or a Fatty Oil.* In a 250-ml Erlenmeyer flask place 10 g of beef tallow, lard, butter, or a vegetable oil, such as corn, olive, peanut, cottonseed, soybean, or safflower oil. Add 5 g of NaOH and 50 ml of 95% ethyl alcohol. Place the flask on a steam bath (or heat in some other way so as to avoid fire hazard from flames) so that the contents become heated to 75-80°, and allow to stand with occasional shaking for 15-20 minutes. Then cautiously pour in 1-2 ml of water and note whether cloudiness develops in the particular part of the solution where the water was added. If any unsaponified fat is still present in the mixture, it will be forced out of solution at this point and produce a characteristic cloudy appearance. In this case the flask should be heated for another 10-15 minutes and tested again with 1-2 ml of water as before. When the saponification is complete, add water to a total of 75 ml and mix the contents of the flask thoroughly. The flask should now contain a perfectly clear, amber-colored, soap solution, which is used for the tests in B, C, and D below.

B. *Separation of Soap by Salting-Out.* To about 10 ml of the soap solution add 10-15 ml of saturated NaCl solution. Let stand until the soap has separated as a curdy, granular mass, and filter. Dissolve the solid in about 10 ml of hot water, and repeat the salting-out process. Dissolve a few particles of the final product in 5 ml of alcohol and test with Hydrion paper (pH range 8-10). Similarly test a sample of commercial soap.

C. *Insoluble Soaps.* To 5-ml portions of the soap solution from A add a few drops of 0.2 M solutions of calcium, magnesium, and lead salts. Repeat with 5-ml portions of a clear, 5-10% solution of a synthetic detergent.

D. *Separation of Equivalent Weight of Fatty Acids.* To the remainder of the soap solution from A add dilute (3 M) sulfuric acid in excess (about 18 ml, test with litmus) and heat at or near the boiling temperature for 2 minutes. Note the oily layer of fatty acids that gradually forms on the surface. Set aside until the fatty acids have come to the top and then filter on a *wet* filter paper, collecting the filtrate in a 400-ml beaker. Save the filtrate for the preparation of glycerol (part E). Wash the fatty acids with hot water two or three times or until freed from sulfuric acid and set aside to see if they will solidify. Compare their physical state with that of fatty acids prepared by other students from different fats.

Accurately weigh 0.300-0.500 g of your fatty acids into a clean, dry 150-ml beaker, dissolve in 25-30 ml of 95% ethyl alcohol, add 3 drops phenolphthalein indicator, and titrate with standard 0.1 N NaOH to the pink end point. Calculate the equivalent weight of the fatty acid sample, referring to Experiment 8, p. 60. The result will be an average value corresponding to the kinds and proportions of the fatty acids present in the triacylglycerols of the original fat. Record your results and compare them with other fats.

E. *Separation and Test for Glycerol.* Neutralize the filtrate from the fatty acids with 1 M NaOH (litmus paper), concentrate first over a flame (under the hood), and finally on a water bath almost to dryness. To the residue add 25 ml of alcohol, warm slightly on a water bath for a few minutes, filter *(turn off burner)*, and test the filtrate for glycerol as described below.

To 1 ml of the above alcoholic filtrate, add 10 drops of 0.1 M sodium periodate ($NaIO_4$), mix, and allow to stand at room temperature for 5 minutes. Then add 10 drops of 10% sodium bisulfite ($NaHSO_3$) solution. Place 8 drops of this mixture in a clean test tube and add 5 ml of chromotropic acid reagent, mix well, and heat for 30 minutes in a boiling water bath. A purple color shows the presence of glycerol. Show the test to your instructor.

Reagents and Materials are listed on page 281.

Name_____ Section _____ Date _____

Experiment 17. SAPONIFICATION OF FATS

A. *Saponification of a Solid Fat or a Fatty Oil*

Name of fat or fatty oil saponified. _____
Name four substances present in the clear solution after saponification in addition to soap, $RCOO^-Na^+$.

_____

Using oleopalmitostearin as an example of a triacylglycerol, write an equation illustrating saponification. Use condensed structural formulas for the fatty acid components. Name the products.

B. *Separation of Soap by Salting-Out*
Describe the physical appearance of your soap.

_____

The approximate pH of your soap._____Of commercial soap. _____

C. *Insoluble Soaps*
Describe what happened when solutions of Ca^{++}, Mg^{++} and Pb^{++} ions are added to
1. Solutions of your soap.

_____

2. Solution of a synthetic liquid detergent.

_____

Write an equation of the reaction of Ca^{++} with a soap, $RCOO^-Na^+$.

_____

D. *Separation and Equivalent Weight of Fatty Acids*
Write an ionic equation for the acidification of a soap, $RCOO^-Na^+$, with dilute H_2SO_4.

_____

Describe the appearance of your fatty acids after filtration and washing free of acid.

_____

Average Equivalent Weight of Fatty Acids:

Weight of beaker and fatty acids   _____

Weight of beaker ....   _____

Weight of fatty acids ....   _____

Ml of NaOH _____ Normality of NaOH _____

Show calculations: (Review calculations in Expt. 8, part C.)

Average equivalent weight   _____

E. *Separation and Test for Glycerol*

How was the glycerol separated from the filtrate from the fatty acids?

_____

_____

Color of your glycerol test.   _____

QUESTIONS

Name_____ Section _____ Date _____

Experiment 17. SAPONIFICATION OF FATS

1. In the saponification of fats, why is ethyl alcohol used as a solvent instead of water?

   _____

   _____

2. What is the chemical function of NaOH?

   _____

3. How was excess NaOH avoided in the separated solid soap?

   _____

4. What happens when ordinary commercial soap is placed in hard water?

   _____

   _____

5. What are the advantages of synthetic detergents compared to ordinary soaps?

   _____

   _____

6. What ingredients would you use to prepare

   a. A *hard* soap?_____

   b. A *soft* soap?_____

7. List the two important factors that influence the melting points of fatty acids.

   (1)_____

   (2)_____

8. Explain why the free fatty acids derived from some soaps solidify, while others remain liquid. Give examples.

   _____

   _____

   _____

9. Name the *main* fatty acids that should be produced from the fat that you saponified in this experiment (consult a textbook or chemistry handbook), and indicate the percentage range (if available) for these fatty acids.

Estimate approximately what the equivalent weight of this mixture of fatty acids should be.

How does this estimated value compare with the results from your titration in part D?

Nonsaponifiable Materials. Sterols and Tocopherols (Vitamin E)

INTRODUCTION

Nonsaponifiable materials are lipid substances that are insoluble in the alkaline aqueous solution after the saponification of a fat as illustrated in the previous experiment. They can be extracted with ethyl ether from the saponification mixture, which contains soap, glycerol, and other polar compounds that are soluble in the aqueous solution. After recovery from the ethereal layer, they can be further examined as to their chemical nature. Nonsaponifiable substances generally consist of fat-soluble vitamins, steroids, pigments, and other miscellaneous compounds. Although they represent only 1-2% of most natural fats, they can be of importance relative to the color, flavor, keeping quality, and nutritional value of fats.

Steroids are compounds derived from a fused, reduced ring structure of three 6-membered rings and one 5-membered ring, and this fused ring structure is referred to as the perhydro-cyclopentano-phenanthrene nucleus (one word, but hyphenated here for your enunciation). *Sterols* are steroids having a hydroxyl group at carbon-3 and an aliphatic side chain of eight or more carbon atoms at carbon-17 of the ring. Cholesterol and ergosterol are important sterols of animal and plant origin, respectively.

The Liebermann-Burchard test is not only a qualitative test for unsaturated sterols, but it is also the basis for the quantitative estimation for cholesterol as described in Experiment 19. The rate of color development is a function of the degree of unsaturation of a particular sterol.

The *tocopherols* (vitamin E) belong to the group of fat-soluble vitamins found associated with lipids when the latter are extracted with a fat solvent. Of the various forms of tocopherols the α-form has the highest biological activity and is found in high concentration in wheat germ oil. The tocopherols are naturally occurring antioxidants which improve the keeping qualities of fats.

PROCEDURE

A. *Liebermann-Burchard Test for Sterols.* Moisture must be carefully avoided in this test, so be certain to use clean, *dry* test tubes.

To 2-ml portions of 0.05% solutions of ergosterol and cholesterol in dry test tubes add 15 drops of acetic anhydride, mix, and add 2 to 5 drops of concentrated sulfuric acid. Mix, and note the color changes. A deep blue green color develops in both tubes, but is preceded in the case of ergosterol by a fleeting red color. Ergosterol develops maximum color in 3 to 5 minutes, while cholesterol requires 18 to 20 minutes.

B. *Modified Furter-Meyer Test for Tocopherols.* To 2 ml of 0.05% solution of α-tocopherol in chloroform add 7 ml of n-butyl alcohol and 1 ml of concentrated nitric acid. Mix and place the tube in a water bath at 80° for 10 minutes. Do *not* exceed this temperature, since higher temperatures may cause ejection of the contents of the tube. What color develops?

Carotenoids and certain steroids produce a yellow to brown color, and when present in sufficient quantity may mask the color due to small amounts of tocopherols.

Reagents and Materials are listed on page 282.

LABORATORY REPORT

Name_____ Section_____ Date_____

Experiment 18.   NONSAPONIFIABLE MATERIALS. STEROLS AND TOCOPHEROLS (VITAMIN E)

A. *Liebermann-Burchard Test*

Describe the color changes. _____

_____

_____

Which sterol changes color more slowly? _____

Why?_____

Considering the chemical nature of the reagents used, why is it essential to use *dry* test tubes?

_____

B. *Modified Furter-Meyer Test for Tocopherols*

Describe the color produced. _____

What substances could interfere with the color of a positive tocopherol test?

_____

Name_____ Section _____ Date _____

Experiment 18. NONSAPONIFIABLE MATERIALS. STEROLS AND TOCOPHEROLS (VITAMIN E)

1. What is meant by nonsaponifiable materials and of what do they consist?

2. When nonsaponifiable materials are extracted by ether, why do soap, glycerol, and ethyl alcohol remain in the alkaline aqueous solution?

_____

3. How would you define each of the following terms?

Steroids _____

Sterols _____

4. Show the structures of cholesterol and ergosterol. Point out the differences in their unsaturation and any other features.

5. What factors hasten oxidative changes in fats?

_____

6. What relationship exists between the keeping quality of a fat and its tocopherol content?

_____

7. Name another vitamin with antioxidant properties. _____

8. What are the names of the compounds abbreviated BHA and BHT and what purpose do they serve as food additives?

BHA _____

BHT _____

Purpose: _____

Determination of Total Cholesterol in Blood Serum

INTRODUCTION

Cholesterol is synthesized in human and animal cells from acetyl coenzyme A (acetyl CoA) into a large 27-carbon steroidal molecule. All of the carbon atoms of cholesterol are derived from the carbon atoms of the acetyl groups of the contributing acetyl CoA molecules. The latter compound is the end product of the beta-oxidation of fatty acids and is also an intermediate product in the oxidation of glucose. Cholesterol, in turn, is a precursor of other physiologically-important compounds, including the steroidal hormones, bile salts, and cholecalciferol (vitamin D_3). In addition, cholesterol appears to be involved in the structural and dynamic properties of cell membranes.

Approximately 75% of the total serum cholesterol exists in the form of cholesteryl esters in which palmitic acid usually is the fatty acid esterified to the hydroxyl group at carbon-3 of the steroidal ring. The remainder of the serum cholesterol is in the free form.

The total serum cholesterol of an adult in the United States roughly falls in the range of 135 to 300 mg per 100 ml of serum, depending on age, sex, and other factors. Various abnormal physiological conditions can cause hypo- and hypercholesterolemia. Atherosclerosis is a condition associated with the accumulation of cholesterol and cholesteryl esters in the inner walls of the arteries, and the build-up of a fatty deposit can cause a heart attack. Elevated levels of serum cholesterol have been related to a predisposition of coronary heart disease and, therefore, the determination of total serum cholesterol is one of the important clinical tests.

This experiment includes two methods for the determination of serum cholesterol: (1) the older Liebermann-Burchard method, and (2) the new enzymatic procedure. The Liebermann-Burchard procedure, in various modifications, has been used for many decades. It is based on the reaction between cholesterol and cholesteryl esters with a reagent mixture of glacial acetic acid, acetic anhydride, and concentrated sulfuric acid (45:45:10 volume ratio) to produce a green solution (see Expt. 18). The student must exercise great care in handling the corrosive color reagent. The amount of cholesterol in the sample can be roughly estimated in this experiment by visually comparing the color intensity of the sample tube with the intensities produced by a series of standard tubes prepared with graded amounts of pure cholesterol. If the instructor so desires, the determination may be done more accurately with a colorimeter or spectrophotometer, but the students must be very careful that the corrosive reagent is not spilled on the expensive instruments. Some interfering substances that can give elevated cholesterol values include: hemolyzed red blood cells, lipoemic serum, and bilirubin.

The new enzymatic method for the determination of cholesterol was described by Allain et al. in 1974. The reagent is an aqueous, buffered mixture of three enzymes: cholesterol esterase, cholesterol oxidase, and peroxidase, plus a chromogenic system of phenol and 4-aminoantipyrine. The esterase catalyzes the hydrolysis of cholesteryl esters in the blood serum to free cholesterol. The cholesterol oxidase catalyzes the oxidation of the secondary alcohol group of cholesterol to a ketone group to yield cholest-4-en-3-one and hydrogen peroxide. In the presence of peroxidase, the hydrogen peroxide is coupled to the 4-aminoantipyrine and phenol to yield a quinoneimine dye that has a maximum absorbance at 500 nm. The color intensity of the dye is directly proportional to the quantity of total cholesterol (free and esterified) in the blood serum sample. These reactions, which are complete in 10 minutes at 37°C at pH 6.7, can be summarized as follows:

$$(1) \quad \text{cholesteryl esters} \xrightarrow[\text{esterase}]{\text{cholesterol}} \text{cholesterol} + \text{fatty acids}$$

$$(2) \quad \text{cholesterol} \xrightarrow[\text{oxidase}]{\text{cholesterol}} \text{cholest-4-en-3-one} + H_2O_2$$

$$(3) \quad 2H_2O_2 + \text{4-aminoantipyrine} + \text{phenol} \xrightarrow{\text{peroxidase}} \text{quinoneimine dye} + 4H_2O$$

The disadvantages of the Liebermann-Burchard method include: (1) the use and handling of a very corrosive acid reagent; (2) the interference by serum proteins unless they are removed by prior precipiation or by the extraction of cholesterol and cholesteryl esters; and (3) higher and more variable cholesterol values. The enzymatic method does not have these disadvantages, since the reactions are carried out in a neutral, aqueous solution and the enzymes are quite specific. The cholesterol oxidase, however, catalyzes the oxidation of 3 β-sterols having a double bond between carbons 5 and 6 or between carbons 4 and 5. Some of these cholesterol derivatives are present in only small amounts, but 7-dehydrocholesterol (vitamin D_3 precursor) occurs normally in serum at a level of 20 mg/100 ml.

PROCEDURE

A. **Blood Serum Sample.** Check with your instructor to find out if you will be furnished with a test tube containing a premeasured volume of an unknown serum sample or if several serum samples will be made available for general class use. If desired, the students may use their own blood for the analysis, and in this case suitable arrangements must be made for taking the blood samples. The freshly-drawn blood is permitted to stand until completely clotted, and then the clot is centrifuged down. The clear serum is carefully poured off and used for the cholesterol determination. The clear serum must not be contaminated with any hemoglobin or the test will be invalid. If animal blood serum is also available, its cholesterol content can be determined and compared with human serum. Hemolyzed red blood cells will interfere in both cholesterol methods given below.

Heparinized blood plasma can be used, but plasma containing oxalate, citrate, fluoride, or EDTA is not recommended, because these reagents give lower cholesterol results.

B. **Liebermann-Burchard Method for the Determination of Total Cholesterol. NOTE:** If your instructor wants you to measure the color intensities of your serum sample and your cholesterol standards with a colorimeter or spectrophotometer, then you will need six, closely matched 18 x 150--mm test tubes. Thoroughly clean and rinse 8-10 of your test tubes (all of same size and brand), fill them with 5-10 ml of distilled water, and dry the outside of the tubes. Using the 18-mm (3/4") test tube holder in the colorimeter, select six test tubes that match the closest at 625 nm. The test tubes must be positioned in the holder with the label *always facing you* because of slight variations in the glass surface. The test tube in the tube holder must be shielded from external light with a paper cup, tin can, or other shield. The selected tubes should check within ± 0.005 absorbance (±1% T). Your instructor will demonstrate how to match the tubes. The use of your matched 18 x 150-mm test tubes for both reaction and color measurement makes it unnecessary to transfer the corrosive reaction mixture into the 13-mm (1/2") cuvettes commonly used with the colorimeter. *If you are not using the colorimeter, but only roughly estimating the cholesterol content by visual comparison with cholesterol standards, you do not have to match the test tubes.*

Label six test tubes: B (Blank), U (Unknown serum sample), numbers 1 through 4 (standards). *Be sure to wear safety glasses* and proceed in accordance with the following steps.

1. Pipet 0.20 ml distilled water into tube B with the 1-ml graduated pipet.
2. Pipet 0.20 ml serum sample into tube U with the 1-ml graduated pipet using a rubber suction bulb.

(**NOTE:** If your instructor provides you with a premeasured 0.20 ml unknown serum sample, use the test tube provided by your instructor for all steps that follow. If you will determine cholesterol with the colorimeter, transfer the reaction mixture to your own matched tube *after step 7.*)

3. Obtain about 4-5 ml of cholesterol standard in a clean, *dry* test tube (2 mg cholesterol/ml in glacial acetic acid). Pipet exactly 0.10, 0.20, 0.30, and 0.40 ml of cholesterol standard into the respective standard tubes labeled 1 through 4. Use the clean, dry 1-ml graduated pipet and a rubber suction bulb. **CAUTION:** Do *NOT* pipet by mouth because the cholesterol is dissolved in glacial acetic acid.

4. Into tubes B and U add 5.0 ml of the color reagent from the 50-ml buret *(Teflon stopcock)* provided for you on the side bench. Add the reagent as follows: Insert the tip of the buret into the test tube *without touching* the wall of the tube. Open the stopcock all the way and allow the first 4.5 ml of color reagent to fall in a rapid, steady stream directly upon the solution in the bottom of the test tube, and then more slowly adding the remainder of the 5 ml of reagent. A minimum of additional mixing will be necessary.

   CAUTION: The color reagent contains concentrated sulfuric acid. *Do not allow it to run down the outside of the test tube or to be smeared around on the desk top.*

5. In the *same manner* as described in step 4, add the volumes of color reagent indicated below into the standard tubes 1-4.

Standard tube no.	1	2	3	4
ml of color reagent	5.10	5.00	4.90	4.80

   (Total volume in *all* tubes is 5.20 ml.)

6. Mix the contents in each tube thoroughly by gentle swirling.

7. Place all tubes in a 37° water bath for 10-15 minutes.

8. Estimation of cholesterol:
   (a) *By Visual Observation.* Compare the color intensity of the unknown serum tube with the color intensities of the standard cholesterol tubes. After completing the table on the report sheet for the graded concentrations of cholesterol in the standard tubes, estimate the amount of cholesterol in the serum sample.
   (b) *By Spectrophotometry (Colorimetry).* If the amount of cholesterol is to be determined more accurately, the color intensities of the solutions are read on the absorbance scale of the spectrophotometer at 625 nm. After the incubation period (step 7), mix by gentle swirling, and wipe all tubes clean and dry. Place the blank tube B into the instrument tube holder and cover the upper part of the projecting tube with a light shield. Set the instrument to zero absorbance (100% T) and read the absorbancies of the serum sample and the cholesterol standards. Review "Notes on the Use of the Spectronic 20" on page 21. (**CAUTION:** Remember all test tubes contain the corrosive acid reagent, including conc. H_2SO_4. Avoid accidental spillage on the instrument or yourself. **WEAR SAFETY GOGGLES AT ALL TIMES!**)

   Record your absorbance values in the table of the report sheet and determine the cholesterol content of your unknown serum sample as instructed there.

C. *Enzymatic Method for the Determination of Total Cholesterol.* To conserve on the expensive enzyme reagent used in this method, your instructor may restrict the use of this reagent. For example, only a few students will set up the reagent blank and cholesterol standard tubes. Each student, however, can be given an unknown serum sample. After color development, the color intensity of your unknown can be measured against one of the several reagent blanks available, and the absorbance value of one of the cholesterol standards will be given to you, or you can measure it yourself. The several reagent blanks and cholesterol standards can be run directly in the matched 1/2″ cuvettes available with the spectrophotometer or colorimeter. Each student can set up his serum sample in a small test tube or culture tube (e.g, 13 x 100 mm) and, after color development, transfer the reaction mixture into a matched cuvette before reading its absorbance. The complete procedure is outlined below, but check with your instructor to determine whether you are to set up a reagent blank or cholesterol standard. Your instructor will demonstrate the proper technique for filling and draining a micropipet.

1. Label three tubes: B (Blank), S (Standard), and U (Unknown serum).
2. To tube B, add 25 microliters (0.025 ml) of water.
3. To tube S, add 25 microliters of cholesterol standard (200 mg/100 ml).
4. To tube U, add 25 microliters serum sample.
5. To each tube, add 3.0 ml of enzyme reagent from the 25-ml buret available on the reagent bench.
6. Cover each tube securely with a piece of Parafilm and mix the contents by inversion several times.
7. Place the tubes in a 37° water bath for 10-15 minutes.
8. Set the spectrophotometer to zero absorbance (100% T) with the reagent blank (tube B), using a wavelength of 500 nm. Review "Notes on the Use of the Spectronic 20" on page 21.
9. Read the absorbance of the cholesterol standard and of the unknown serum. Readings should be completed within 30 minutes. (**NOTE:** If the solution to be read is transferred to a matched cuvette, be sure that the inside of the cuvette is clean and *dry*.)
10. Record the absorbance readings on the report sheet and calculate the total cholesterol content of the unknown serum in mg/100 ml.

Note to the Instructor: A volume of about 3 ml is required to cover the light path in the Spectronic 20 when using the 12 mm i.d. (1/2") cuvette. If microcuvettes and adapter are used, it is possible to use smaller volumes of enzyme reagent to reduce cost. A volume ratio of 10 μl serum (standards and blank) to 1.0 ml of enzyme reagent is recommended.

REFERENCES

ALLAIN, C. A.; POON, L. S.; CHAN, C.S.G.; RICHMOND, W.; and FU, P. C. 1974. *Clin. Chem.* 20:470.
Cholesterol (Liebermann-Burchard method). 1979. *Tech. Bull. CH 172-5603.* Miami, Fla.: Dade Division Amer. Hosp. Supply Corp.
HUANG, T.; CHEN, C.; WEFLER, V.; and RAFERY, A. 1961. *Anal. Chem.* 33: 1405.
TAYLOR, R. P.; BROCCOLI, A. V.; and GRISHAM, C. M. 1978. *J. Chem. Educ.* 55:63.
Total cholesterol. 1979. *Tech. Bull. No. 350.* St. Louis, Mo.: Sigma Chem. Co.

Reagents, Materials, and Equipment are listed on page 282.

LABORATORY REPORT

Name_____ Section _____ Date _____

Experiment 19. DETERMINATION OF TOTAL CHOLESTEROL IN BLOOD SERUM

A. *Blood Serum Sample*

Sample No. _____

Sample information, if any_____

B. *Lieberman-Burchard Method for the Determination of Total Cholesterol*

By comparison with the standard cholesterol tubes, the milligrams of cholesterol in 0.20 ml of blood serum is determined. By what factor must this result be multiplied to express the cholesterol content as mg per 100 ml serum? Show work.

The factor is_____

Complete the table below by calculating the mg of cholesterol in each of the standard tubes, and then multiply these values by the above "factor" to express the results as mg cholesterol per 100 ml serum. If the cholesterol in the serum was estimated by visually comparing the sample tube with the standard tubes, enter your estimated value in the last column.

If the cholesterol was determined by spectrophotometry, record the absorbancies of the standards and the serum sample in the table below. Prepare a standard curve by plotting the absorbancy of the cholesterol standards on the vertical axis (ordinate) against the cholesterol content of each tube expressed as mg cholesterol per 100 ml serum (from table) on the horizontal axis (abscissa). Using the absorbancy of the serum sample, read off from the standard curve the cholesterol content of the serum sample. Record your values in the table.

	Cholesterol standards (2 mg/ml)				*Unknown*
Tube Number	*1*	*2*	*3*	*4*	*U*
mg cholesterol per tube					-----
mg cholesterol per 100 ml serum					
absorbance readings (if taken)					

C. *Enzymatic Method for the Determination of Total Cholesterol*

Volume: Unknown serum and cholesterol standard _____ μl

Absorbance: Unknown, A_u = _____   Standard, A_s = _____

Serum Cholesterol = $\dfrac{A_u}{A_s} \times 200$, where 200 = concentration (mg/100ml) of cholesterol standard

Show your calculations:

QUESTIONS

Name_____ Section _____ Date _____

Experiment 19.   DETERMINATION OF TOTAL CHOLESTEROL IN BLOOD SERUM

Answer *all* questions regardless of cholesterol method used.

1. a. Approximately how much of the total cholesterol in serum is present as cholesteryl esters?

   _____

   b. What is the normal range of total serum cholesterol in an adult in the United States?

   _____

   c. What is the usual serum cholesterol level for a person of your age and sex in this country?

   _____

2. Does all of the cholesterol in our body come from dietary intake? Explain.

   _____

   _____

   _____

3. Give 4 or 5 examples of steroids in our body that are synthesized from cholesterol.

   _____

4. *Give 2 different human* pathological conditions in which cholesterol is involved. Comment briefly.

   a. _____

   _____

   b. _____

   _____

5. What is the chemical composition of the Liebermann-Burchard reagent?

   _____

6. Name several interfering substances in each of the two cholesterol methods.

   Liebermann-Burchard _____

   Enzymatic _____

7. What is the main disadvantage or problem in using each method?

   Liebermann-Burchard _____

   Enzymatic _____

8. List the 3 enzymes involved in the enzymatic method for determining serum cholesterol and give the function of each.

_____

_____

_____

9. Write the structures of the following:
   a. Cholesterol palmitate   b. Cholest-4-en-3-one

10. List the reagents which produce the quinoneimine dye in the enzymatic method.

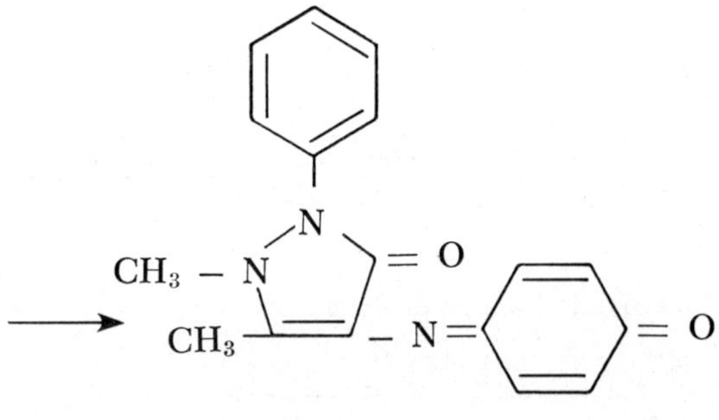

quinoneimine dye

_____

Unsaturation of Fats. Iodine Number

INTRODUCTION

Each carbon-to-carbon double bond of an unsaturated fat under suitable conditions readily combines with two atoms of halogen (iodine, bromine, and chlorine). The number of grams of iodine absorbed by 100 grams of fat, called the "iodine number," therefore is a measure of the degree of unsaturation. Two methods are generally used for iodine number determination: the Wijs method, which uses iodine chloride (ICl); and the Hanus method, which uses iodine bromide (IBr). The two halogens in each method are present in equivalent amounts, but the result is calculated as if only iodine was present.

The Hanus reagent, which is used in this experiment, is more stable, but the Wijs method gives results 2 to 5 percent higher and the iodine numbers so obtained are closer to the theoretical values.

To determine the iodine number a weighed sample of fat is allowed to react with excess iodine bromide and the amount of unreacted IBr is then measured by adding potassium iodide and titrating the liberated iodine with standard sodium thiosulfate solution. The process may be summarized by the following equations:

$$I_2 + Br_2 \rightleftarrows 2IBr \text{ (iodine bromide)}$$
$$Fat + \text{excess } IBr \longrightarrow \text{Halogenated fat} + \text{unreacted } IBr$$
$$KI + IBr \longrightarrow I_2 + KBr$$
$$I_2 + 2Na_2S_2O_3 \longrightarrow 2NaI + Na_2S_4O_6$$

The iodine number of a particular fat or fatty oil is never reported in the literature as a fixed value, but rather as a numerical range, since the degree of unsaturation of that fat or fatty oil may vary due to the conditions under which the lipid was produced. For example, the iodine number of butter is given as 25-40 because the butterfat composition may depend on the breed or diet of the animal. This holds true for all fats and fatty oils from animals and plants. Generally, solid animal fats will have an iodine number below 70, while the vegetable oils are above 75. The iodine number is related to the percentages of saturated and unsaturated fatty acids in the triacylglycerols of the particular fat or fatty oil. In solid fats palmitic and stearic acids predominate, with oleic acids mainly responsible for the unsaturation. In fatty oils, however, the percentages of saturated fatty acids are significantly lower, while those of oleic and linoleic acids increase markedly until the iodine value approaches 200, at which point linolenic acid predominates (e.g., linseed oil).

Coconut oil is a vegetable oil having the very low iodine number of 8-10. It owes its oily nature to the high percentages of short-chain saturated fatty acids of the C_6 to C_{14} variety. Only small amounts of palmitic, stearic, and oleic acids are present.

PROCEDURE

A. *Sample Analysis.* Each student will be given a sample of an unknown fatty oil in a small test tube. Obtain a long-tip, disposable pipet and a rubber bulb from the stockroom or reagent shelf.

Clean and *dry* two 250-ml Erlenmeyer flasks. (Dry inside of flask with acetone and air.) Place the numbered flask on the *analytical* balance pan (ARREST position) and weigh in the *partial* release position, and finally in the *complete* release position. Record weight to the closest 0.0001 gram. *Before* removing flask, be sure the balance is in ARREST position to prevent jarring and damaging balance mechanism.

Draw a portion of your unknown fatty oil into the disposable pipet and transfer about 0.2 g (13-14 drops) of the oil into the Erlenmeyer flask. Return the flask to the balance and weigh again, using appropriate care as outlined above. Record weighings on report sheet. Repeat the procedure with the second numbered flask. Return the rubber bulb when you have finished weighing the duplicate samples, but keep the disposable pipet until you have satisfactorily completed the entire experiment and then discard the pipet.

Add 10 ml of chloroform to each flask and dissolve the sample by swirling gently. Stopper the flasks with clean #6 rubber stoppers. To each flask add exactly 25.0 ml of Hanus iodine solution from the automatic-filling pipet or buret provided. Close the flasks with the rubber stoppers, mix well by swirling, and allow them to stand at room temperature for 30 minutes in a dark cabinet with occasional swirling. (CAUTION: Hanus iodine solution is a corrosive solution of iodine and bromine in glacial acetic acid. If the automatic-filling pipet or buret is not available, carefully use a 25-ml pipet with bulb suction. DO NOT use mouth suction to fill the pipet with this very corrosive reagent).

Add 10 ml of 15% KI solution, mix thoroughly and add about 50 ml of water, washing down any iodine solution on the wall of the flask or on the stopper. Titrate with 0.10 N sodium thiosulfate solution from a 50-ml buret until the yellow color of the solution has almost disappeared. With the addition of 2 ml of 1% starch the solution turns a deep blue. Continue the titration until the blue starch-iodine color disappears, mixing well during the final stages of titration. To ensure complete removal of the iodine from the chloroform layer, stopper the flask and shake vigorously. If the blue starch-iodine color returns, complete the titration. Record sample weights and titrations on the report sheet. The instructor should demonstrate the titration procedure by running a blank titration as described below.

Blank titration. Check with your instructor whether you will be given the blank titration value for the Hanus iodine solution, or if you are to run a blank yourself, depending on the supply of Hanus solution. In either event the procedure is as follows: To a clean, dry 250-ml Erlenmeyer flask add 10 ml of chloroform, 25.0 ml of Hanus reagent, 10 ml of 15% KI solution and 50 ml of water. Mix and titrate the iodine with standard thiosulfate solution as described above. The 30-minute standing time can be omitted. The ml of thiosulfate required is equivalent to the total halogen content of the Hanus solution.

B. *Calculations.*

The milliequivalents (meq) of iodine *absorbed* by the sample are:

$$\text{meq of I} = (B-S) \times N$$

where B = ml $Na_2S_2O_3$ for blank titration

S = ml $Na_2S_2O_3$ for halogen *unabsorbed* by sample

$(B-S)$ = ml $Na_2S_2O_3$ equivalent to halogen *absorbed* by sample

N = normality of $Na_2S_2O_3$

In the reaction with thiosulfate, $I_2 + 2S_2O_3^= \rightarrow 2I^- + S_4O_6^=$, one atom of I gains 1 electron. Therefore, the gram-equivalent weight (eq wt) of iodine is:

$$\text{eq wt I} = \frac{\text{atomic wt of I}}{\text{electrons gained/atom}} = \frac{126.9}{1} = 126.9 \text{ g}$$

$$\text{meq wt I} = 126.9 \text{ g} \div 1000 = 0.1269 \text{ g}$$

The weight of iodine absorbed by the sample is:

$$\text{grams I absorbed} = \text{meq I absorbed} \times \text{meq wt I}$$
$$= (B-S) \times N \times 0.1269 \text{ g}$$

Since the Iodine Number equals the grams iodine absorbed by 100 g of fat, then

$$\frac{(B-S) \times N \times 0.1269 \text{ g}}{\text{sample wt in grams}} = \frac{\text{g Iodine}}{100 \text{ g fat}}$$

$$\text{g Iodine} = \frac{(B-S) \times N \times 0.1269 \text{ g} \times 100 \text{ g}}{\text{sample wt in grams}}$$

or,

$$\text{Iodine No.} = \frac{(B-S) \times N \times 12.69 \text{ g}}{\text{sample wt in grams}}$$

REFERENCE

POMERANZ, Y. P., and MELOAN, C. E. 1978. *Food analysis: theory and practice*. Rev. ed. Westport, Conn. The AVI Publishing Co.

Reagents and Materials are listed on page 283.

Name_____ **Section**_____ **Date**_____

Experiment 20. UNSATURATION OF FATS. IODINE NUMBER

Unknown Sample No. _____

	Sample 1	Sample 2
Weighing data:		
Fatty oil + flask	_____	_____
Erlenmeyer flask	_____	_____
Sample weight, grams	_____	_____
Titration data:		
Normality of $Na_2S_2O_3$ _____ N		
ml $Na_2S_2O_3$, blank (B)	_____	_____
ml $Na_2S_2O_3$, excess iodine (S)	_____	_____
ml $Na_2S_2O_3$, absorbed iodine (B—S)	_____	_____

Calculations: (Show calculations)

Iodine Number of Unknown: .... _____   _____

Average Iodine No. _____

Name _____ Section _____ Date _____

Experiment 20.   UNSATURATION OF FATS. IODINE NUMBER

1. Explain the relationship between the iodine number and the fatty acid composition of most fats.

   _____

   _____

2. How do you account for the following:
   a. The low iodine number (8-10) of coconut oil?

   _____

   _____

   b. The high iodine number (175-205) of linseed oil?

   _____

   _____

3. Indicate the range of the iodine numbers of animal fats as compared to those of vegetable oils.

   _____

4. a. What is meant by a "polyunsaturated fat"?

   _____

   b. Can you tell from the iodine number if a fat is polyunsaturated? Explain.

   _____

5. a. How many *atoms* of iodine would be absorbed by one molecule of each of the following triacylglycerols?

   Tristearin _____ Triolein _____ Trilinolein _____ Trilinolenin _____
   b. Calculate the iodine number of tristearin, triolein, and trilinolenin if their molecular weights are 891, 885, and 873, respectively. Atomic wt of $I = 126.9$. Show work.

6. Calculate the iodine number of linoleopalmitolino!enin whose molecular weight is 881. Show work. Ans. 144.

7. A 0.250 g sample of butter was dissolved in $CHCl_3$ and treated with 25.00 ml of Hanus iodine solution. After 30 minutes KI was added and the liberated iodine required 40.15 ml of 0.104 N $Na_2S_2O_3$ solution. The blank titration for 25.00 ml of Hanus solution was 46.50 ml of the standard $Na_2S_2O_3$. Calculate the iodine number of the butter. Show work. Ans. 33.5.

Thin-Layer Chromatography of Lipids

INTRODUCTION

Thin-layer chromatography and paper chromatography offer a means of separating small amounts of various mixtures into their individual constituents, provided a suitable solvent system can be found. Originally, mixtures of plant pigments were separated into their components by column chromatography in which the sample, dissolved in a suitable solvent (e.g., petroleum ether) was passed through a column of powdered solid adsorbent, such as calcium carbonate, sucrose, alumina, or other materials. Because the mixtures separated into bands of colored substances, the process was called "Chromatography" (Greek chroma = color) even though today the same term applies to the separation of colorless compounds whose presence can be detected by a variety of techniques. Thin-layer chromatography (TLC) and paper chromatography may be considered as methods analogous to an open column.

Thin-layer chromatography utilizes a sorbent coating, such as silica gel (SiO_2) or alumina (Al_2O_3), on a glass plate or a flexible plastic sheet. In this experiment, three of the most important classes of lipids, namely triacylglycerols, steroids, and phospholipids, can be distinctly separated on a silica gel coated support medium. The silica gel is activated by heating in an oven at 100°C for 15-30 minutes. After activation by heat, the hydrated silica gel still contains about 10% water. The sample is applied to the activated silica gel by spotting with a capillary pipet. The flexible sheet or the glass plate is placed out of contact with the solvent system in a closed container to permit equilibration between the vapor and liquid states of the solvent mixture. Equilibration is important so that there is no change in the composition of the solvent mixture, the *mobile phase*, as it moves up the coating when the plate is allowed to come in contact with the solvent system. As the mobile phase rises through the silica gel by capillary action (the *development* of the chromatogram), it carries along the individual components of the sample at different rates, so that they become physically separated on the coated plate or sheet. The rate of travel of the individual components as the mobile phase flows over the stationary phase depends on the distribution of each component between the two phases. The distribution of a component of an applied sample between the two phases during the development of a chromatogram depends on whether the distribution process involves an *adsorption* or a *partition* mechanism, or a combination of both mechanisms.

In the *adsorption* process, the *silica gel* (or other adsorbent) is the *stationary* phase, and each component of the sample distributes itself between the active sites on the stationary phase and the mobile phase as it migrates up the adsorbent coating. This adsorption mechanism predominates when separating *lipophilic* substances with low or medium polarity, as is the case in this experiment. In addition, relatively nonpolar solvents are used (e.g., low-boiling hydrocarbons, alkyl halides, and ether).

The *partition* mechanism predominates when the components of an applied sample are *hydrophilic* (water-loving and polar), and the solvents in the mobile phase are polar (e.g., alcohols, acetone, and water). In this respect, partition TLC is similar to the separation of polar substances by paper chromatography (Expt. 25). In both chromatographic methods, *water* is the *stationary* phase: the water in the hydrated silica gel in TLC, and the water bound to the cellulose fibers in paper chromatography. The relative solubility of a hydrophilic component in the stationary and mobile phases is referred to as the *distribution coefficient*, K, sometimes called the partition coefficient.

$$K = \frac{\text{conc. of solute in stationary phase (water)}}{\text{conc. of solute in mobile phase}}$$

If the distribution coefficients of the components of a sample mixture are sufficiently different in a particular solvent system, they can be separated successfully.

Before the *solvent front* reaches the top edge of the plate, the development is discontinued. After the solvent has evaporated from the coating, visualization of the component spots must be made if they are not themselves colored. This can be accomplished in a variety of ways. In the case of lipids used in this experiment, visualization can be done (1) by exposure of the plate to iodine vapors; (2) by spraying with alkaline permanganate solution; (3) with fluorescent indicators and UV light; or (4) by spraying with sulfuric acid and charring the spots by heating. Once the spots have been located, the R_f values can be calculated as discussed in a later section.

Although thin-layer and paper chromatography are similar in principle, TLC has a number of advantages over paper that have made TLC more widely used: (1) TLC is more rapid, generally requiring only 20-40 minutes for the development step; (2) the method is more sensitive, in some cases levels as low as 10^{-9} gram of a substance have been detected; (3) more reactive reagents can be used both in the solvent systems and in visualization techniques (e.g., charring spots with sulfuric acid); (4) the method can be scaled up directly to column chromatography; and (5) experimental conditions, such as type of adsorbent, activation time, and temperature, can be varied readily.

PROCEDURE

A. *Preparation of Silica Gel Chromatograms for Sample Application.* You will use either a commercial precoated sheet of silica gel or you will coat glass plates with silica gel. Therefore, you will follow only *one* of the two procedures given in this part of the experiment, unless your instructor wants you to use your prepared plates as well as commercial precoated sheets for comparative purposes.

(1) *Using Commercial Precoated Silica Gel Sheets.* A 100-micron-thick layer of silica gel, containing a polymeric binder (e.g., polyacrylic acid), is coated on a flexible support that is not affected by the iodine vapor used in this experiment for visualization of the spots after development. The sheets are available with a fluorescent indicator for visualization under ultraviolet light.

From your instructor obtain a chromatographic sheet that has been precut to 50 x 95 mm (2" x 3¾"). Place it in an oven at 100°C for 15 to 30 minutes to activate the silica gel. (Silica gel readily adsorbs moisture and other vapors from the atmosphere. Chromatograms should be activated before use when lipophilic substances are chromatographed. If they are not used shortly after activation, they may be stored in a desiccator.)

Using figure 21.2 as a measuring guide, lightly draw a base line with a soft pencil 18 mm from the *bottom* edge of the chromatogram and a short mark 60 mm *above* the base line. Make a *light*, short, vertical pencil mark at the center of the base line and two light marks to the left and two marks to the right of the center with 8 mm between the marks. (Total of 5 marks.) Starting at the left end of the base line, identify each mark with L for lecithin, C for cholesterol, O for vegetable oil, M for mixture, and U for unknown. These are the points of sample application later. Identify the chromatogram with your initials at the upper edge.

(2) *Coating Glass Plates with Silica Gel.* Lay four clean 40 x 90-mm glass plates near the top edge of a paper towel on a flat, level surface so that they are butted together as shown in figure 21.1. The paper towel will absorb the excess slurry of coating material. From a roll of adhesive tape (cloth type) tear a

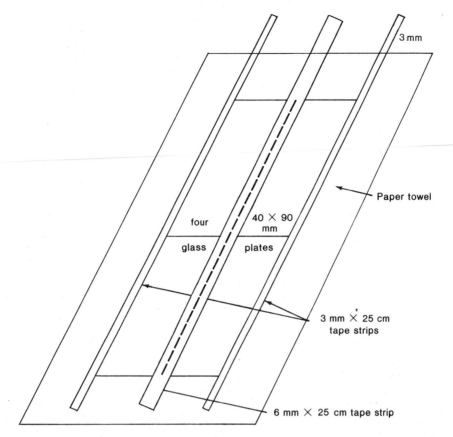

Figure 21.1. Lay-out of glass plates for coating with silica gel.

strip about 6 mm wide by 25 cm long. Place the strip so that the tape straddles the longer edges between the plates, allowing 4 cm of tape to extend beyond the top edge of the plates and adhere to the bench top. Tear two more strips of tape 3 mm wide by 25 cm long and place each tape on the face of the glass plate along the longer outside edges, again allowing them to adhere to the bench top. These tapes will permit the application of a uniform coating of adsorbent about 0.25 mm thick. Wipe the surface of the glass plates with a tissue wetted with acetone to remove the last traces of oil and dirt.

Weigh 3 grams of silica gel (containing 14% $CaSO_4 \cdot \frac{1}{2}H_2O$ as a binder) and transfer it to a 50-ml Erlenmeyer flask fitted with a rubber stopper. Add 6 ml of 3% starch solution, stopper the flask, and shake vigorously for 30 seconds. Immediately pour the slurry along the upper edge of the glass plates between the tapes. Hold a scoopula across the top edge of the glass plates with the convex face of the scoopula toward you, and allow its long edge to rest on the strips of tape. Draw the scoopula smoothly toward you while firmly, but lightly, pressing it against the tapes. You may repeat the process of drawing the scoopula from the top to the lower edges of the plates once more. Do not go back and forth. Complete the spreading of the slurry as quickly as possible, since the $CaSO_4$ will begin to set in a few minutes. Air dry 30 minutes or longer. Plates can be prepared during one lab period and used in the next.

Carefully remove the tapes from the dry glass plates and wipe off any silica gel on the backs or edges. Heat the plates in an oven for one hour at 100°C, to remove moisture and to activate the sorbent. The film should be hard enough to mark lightly with a soft pencil. The starch improves the surface hardness of the coating.

Using two of your best plates and a soft pencil, lightly draw base lines 18 mm from the *bottom* edges and short marks 60 mm *above* these base lines as shown in figure 21.2.

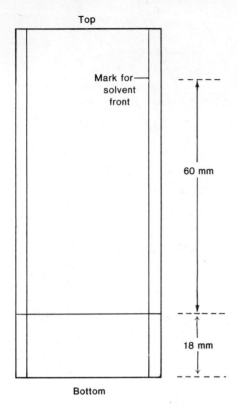

Top

Mark for solvent front

60 mm

18 mm

Bottom

Figure 21.2. Guide for marking chromatogram.

Place the plate over the illustration and use it as a guide. Avoid rubbing the coating with your hand. In drawing the light base line across the coating, do not disturb the silica gel. If sorbent is removed and the glass is exposed, *the developing solvent cannot bridge the gap and the plate is ruined.* Make a *light,* short vertical pencil mark 5 mm from the left end of the base line and then make three additional marks with 7-mm spacings between them. (Total 4 marks.) On one glass plate lightly write below each mark the letters L, C, O, M, respectively; on a second plate write C,O,U,M, respectively. (L = lecithin, C = cholesterol, O = vegetable oil, M = mixture of the three lipids, and U = unknown.) Identify your plates at the top edge with your initials in pencil.

B. *Spotting the Lipid Samples.* Your instructor will demonstrate the proper spotting technique. If you are using the commercial precoated chromatograms, all samples will be spotted on one sheet that has been marked in accordance with part A (1). However, if you are spotting self-prepared glass plates, use two plates as described in part A (2) to prevent sample crowding. The samples are lecithin, cholesterol, vegetable oil, and a mixture of these three lipids, plus an unknown, all of which are at a concentration of 5 mg/ml of lipid in chloroform. They will be available in screwcap vials together with separate 1-microliter (1-μl) capillary pipets for spotting the lipid samples. Insert the pipet into the holder provided and dip it into the sample solution; the pipet is self-filling by capillary action. Gently touch the tip of the pipet to the appropriate mark on the base line. Since the pipet is self-draining, the sample solution will be transferred to the silica gel where the chloroform evaporates. Practice filling and draining the pipet on a filter paper before spotting your chromatograms. Pipets should be cleaned with chloroform by filling and draining them 4-5 times on a filter paper.

C. *Developing the Chromatogram.* Prepare a chromatographic chamber using a clean, dry 250-ml beaker fitted with an aluminum foil cover that does not extend more than 12 mm (½″) down the side of the beaker. Initially the beaker will stand tilted at an angle, using the edge of the base of a ringstand or some other object.

Put 10 ml (graduated cylinder) of the solvent system in the 250-ml beaker. The solvent system is a petroleum ether, ethyl ether, and glacial acetic acid in a volume ratio of 70:30:1. (*Flammable!* Keep away from flames.) *Without delay,* to prevent evaporation, set the beaker with one side slightly elevated as described above. Place a chromatogram in the beaker *without* its bottom edge touching the liquid. *Immediately* cover the beaker with the Al foil, carefully sealing the foil around the top of the beaker without splashing liquid on the chromatogram. Allow the whole system to stand tilted for 5 minutes to equilibrate. Then move the beaker to an upright position so that the liquid (mobile phase) contacts the bottom of the chromatogram. Wait until the solvent rises to the upper pencil mark 60 mm above the base line (about 10-14 minutes). Remove the chromatogram from the beaker and allow the solvent to evaporate at room temperature. (**NOTE:** Should the solvent front accidentally go above the upper pencil mark, a new mark must be made immediately when the chromatogram is removed, before solvent evaporates.) Discard the flammable solvent in an approved manner, do not reuse it.

D. *Visualization of Spots with Iodine Vapor.* Place the chromatogram in a 250-ml beaker containing several crystals of iodine (hood) and cover the beaker with a watch glass. After exposure to iodine vapors for 30-45 minutes, the lipids are located as brown spots on a white to yellow background. The color intensity of the spots depends on the concentration of the lipids. Remove the chromatogram from the beaker and expose it to the atmosphere for at least 15 minutes to allow iodine to vaporize from the silica gel coating. Encircle the spots with a pencil, since they will gradually disappear. They usually reappear if the chromatogram is again exposed to iodine vapor.

E. **R$_f$ Calculations. Identification of Unknown.** Measure and record the distance (in mm) from the base line to the center of each spot. Calculate the R$_f$ value of each lipid and identify your unknown.

$$R_f = \frac{\text{distance spot moved}}{\text{distance solvent moved}}$$

If you used a commercial precoated chromatogram, tape it to your report sheet with transparent tape. If you used glass plates, they must be washed and *returned* to your instructor or stockroom.

F. **Separation of Other Materials.** A mixture of carotenes from carrots prepared as described in Experiment 31 can be chromatographed by TLC, or Experiments 21 and 31 can be done simultaneously. Materials extracted from leafy plants, such as lettuce or spinach, can also be tried.

REFERENCES

STAHL, E., ed. 1969. *Thin-layer chromatography.* 2d ed. New York: Springer-Verlag, pp. 376-79.

UMBREIT, W. W.; BURRIS, R. H.; and STAUFFER, J. F. 1972. *Manometric and biochemical techniques.* 5th ed., chap 15. Minneapolis, Minn.: Burgess Publishing Co.

Thin-layer chromatography (TLC). 1979 *Eastman Dataservice Catalog,* Rochester, N.Y.: Eastman Kodak Co.

Reagents and Materials are listed on page 284.

LABORATORY REPORT

Name_____ Section_____ Date_____

Experiment 21. THIN-LAYER CHROMATOGRAPHY OF LIPIDS

Composition of Solvent System_____

	— Chromatogram 1 —		— Chromatogram 2 —	
	mm	R_f value	mm	R_f value
Solvent front	_____	__1.00__	_____	__1.00__
Lecithin	_____	_____	_____	_____
Cholesterol	_____	_____	_____	_____
Vegetable oil	_____	_____	_____	_____
Mixture of above ...... (3 spots)	_____	_____	_____	_____
	_____	_____	_____	_____
	_____	_____	_____	_____
Unknown ...., (1 to 3 spots)	_____	_____	_____	_____
	_____	_____	_____	_____
	_____	_____	_____	_____

Unknown No. _____ My unknown is _____

NOTE: *Chromatogram 2* column is for the second glass plate or a possible duplicate commercial precoated sheet.

Data for other chromatographed materials, if any:

Name_____ Section _____ Date_____

Experiment 21.   THIN-LAYER CHROMATOGRAPHY OF LIPIDS

1. What quantity of lipid is present in the volumes given below if a 0.05% solution of the lipid in chloroform is used?

_____ g per 100 ml   _____ mg per μl

_____ mg per ml   _____ μg per μl

2. Distinguish between adsorption chromatography and partition chromatography.

_____

_____

_____

3. What type of solvent is used in adsorption and partition chromatography? Name several solvents used in each.

Adsorption _____

Partition _____

4. What general type of substances are separated by

Adsorption chromatography? _____ Partition chromatography? _____

5. Considering the polarity of the solvent system, account for the order of migration of the lipid samples used in this experiment.

_____

_____

6. Would you expect amino acids or carbohydrates to separate with the same solvent system as used for lipids? Explain and indicate what changes you would make, if any.

_____

_____

_____

Determination of Fat by Soxhlet Extraction

INTRODUCTION

Lipids are generally defined as that group of compounds that can be extracted from plant and animal tissues by certain organic solvents, such as ethyl ether, chloroform, carbon tetrachloride, or petroleum ether. Based on this solubility classification, lipids include fats (triacylglycerols), waxes, phospholipids (lecithin, etc.), terpenoids (e.g., steroids, carotenoids), fat-soluble vitamins, and many other substances soluble in the above solvents. Since fats are so widely and abundantly distributed in nature, the extraction of biological materials and foodstuffs by these "fat solvents" has been simply referred to as the "Determination of Fat." Proteins and carbohydrates are insoluble; hence this extraction procedure is an excellent means of separating lipids from these two major groups of biological materials.

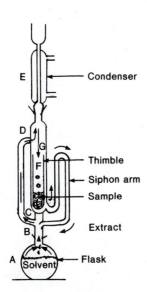

Figure 22.1. Soxhlet Extractor. The flat-bottom flask (A) is heated on an electric hot plate. An extraction chamber with a stopcock drain is recommended. (From *Food Analysis: Theory and Practice* by Y. Pomeranz and C. E. Meloan. Copyright 1978 by the AVI Publishing Company. By permission of the publisher.)

The Soxhlet method is one of the extraction procedures commonly used (fig. 22.1). During the extraction process, the vapors from the boiling solvent in the heated flask (A) rise through the glass tubing at (B) since they cannot enter the extraction chamber, which is sealed off at the bottom (C). The rising vapors reenter the chamber at (D) and are condensed by the water-cooled condenser (E). The condensed solvent fills the extraction chamber (F) and dissolves the fat from the sample in the paper thimble standing in the chamber. A siphon tube connected to the bottom of the chamber also fills with solvent, as the solvent level rises in the chamber until the level reaches a point (G) at which the siphon automatically starts and drains the solvent and the extracted fat from the chamber into the flask (A) below. The extraction cycle repeats itself about every 10-15 minutes, and after 16-20 hours all of the fat has been removed from the sample and transferred to the extraction flask. The analysis is then completed as described in the procedure below.

A wide variety of biological materials of foodstuffs could be suggested for a fat determination. However, some of the cereals adapt themselves more readily for illustrating the Soxhlet extraction; they have a relatively low moisture content, can be easily weighed directly into the paper thimble and may be readily dried without further handling. In any event the sample should be crushed, shredded, or ground, and should be dried at 105°C for 1-2 hours before solvent extraction. Removal of water facilitates fat extraction, reduces the tendency of transferring water with the solvent during siphoning, and also reduces difficulties in bringing the extracted fat to a constant weight in the final steps of the analysis. If the sample has a high moisture content, it must be weighed and dried in a dish before quantitatively transferring it to the thimble and, if it becomes lumpy as a result of drying, it must be ground before transferring.

PROCEDURE

Instructor's Note: This experiment can be readily *demonstrated* to the class with a single extraction asembly if the cost of multiple assemblies and heaters are prohibitive. Students can be given hypothetical weighing data and asked to complete the Laboratory Report and Questions. If only a limited amount of equipment is available, students can set up a single sample instead of the duplicate samples indicated below.

A. *Preparation and Extraction of Sample.* The apparatus consists of a Soxhlet extractor, a paper thimble, and an extraction flask. This assembly will be set on an electric heater and attached to a reflux condenser. Since the thimbles are used repeatedly, clean out any material that may be in them from previous use. The clean, numbered flasks are heated in the oven at 105°C for at least thirty minutes, cooled in the desiccator, and weighed on the analytical balance. Record all weights in your *laboratory manual, not* on a piece of scrap paper. Thimbles should be numbered same as flask.

The instructor will furnish you with a cereal or a blend of cereals. Weigh two 6-gram samples into thimbles using the *analytical balance.* For convenience, set the thimble in a clean 100-ml beaker. Record the weights of the thimble and beaker. Add the sample to the thimble and again record the weight accurately. The difference in weights represents the weight of the sample. Place a *small* wad of glass wool loosely in the thimble on top of the sample. This prevents the sample from floating out of the thimble during the extraction process. Place the two thimbles in a 250-ml beaker in the oven and dry for 1-2 hours at 105° to remove most of the moisture.

Before the end of the lab period, remove the thimbles from the oven and insert them into the extractor chambers. Insert the extractor into the numbered extraction flask containing 90 ml of chloroform (use graduate provided). This volume of chloroform will ensure an adequate amount of solvent remaining in the extraction flask when the extractor chamber is filled just prior to siphoning, thus preventing scorching of the extracted fat in the extraction flask. Turn on the heater after the extraction assembly has been attached to the condenser and extract for 20 hours. If the extraction chamber is provided with a stopcock at the bottom of the chamber for solvent drainage later, *be sure that the stopcock is closed.* Check that all glass joints are tight.

At the next laboratory period the procedure for the removal of the chloroform depends on whether the extraction chamber has a stopcock drain. Use one of the following two methods:

(1) If the extraction chamber *has* a stopcock, recover most of the chloroform from the extractor by opening the stopcock and draining the solvent into the "Used Chloroform" bottle. Continue heating and draining the chloroform before it siphons over until a small amount remains in the extraction flask. Then turn off the heater, remove the apparatus from the condenser, disconnect the extractor from the flask and place the extractor containing the thimble under the hood. Heat the flask on a hot plate under the hood until most of the chloroform has evaporated. *Do not leave the flask unattended,* since it may go to complete dryness and scorch the fat, thereby yielding low percentage of fat.

(2) If the extraction chamber *does not* have a stopcock for draining the chloroform, turn off the heater after siphoning of chloroform has occurred and allow the chamber to collect chloroform below the siphoning points. After the flask has cooled somewhat, disconnect the chamber from the condenser and, under the hood, carefully remove the extraction chamber from the flask. Pour the chloroform *from the chamber* into the "Used Chloroform" bottle, but hold the chamber properly so that the chloroform does not accident-

ally cause siphoning to occur. The chloroform *in the extraction flask* is recovered using a distillation set-up or a rotary evaporator. If the flask is heated on a hot plate or heating mantle when distilling chloroform, do not allow the flask to go to complete dryness, otherwise you might scorch the fat.

Now place the extraction flask containing the fat in the oven at 105° for 30-45 minutes, cool in the desiccator, and weigh. Repeat the oven-heating, cooling, and weighing until the difference between two *successive* weighings is not more than *0.0025 grams.* (What does the difference of 0.0025 grams represent in terms of % of fat from a 5-gram sample?)

B. *Calculations.* Calculate percentages of fat in each sample to two decimal places and then average your results.

$$\% \text{ Fat} = \frac{\text{Weight of flask and fat} - \text{Weight of flask}}{\text{Weight of sample}} \times 100$$

The proximate percentages of fat in some common foods are shown in the table given in conjunction with question 5. A very extensive compilation will be found in the two Government publications listed in References.

REFERENCES

ADAMS, C. F. 1975 *Nutritive value of American foods.* U.S. Dept. of Agr. Handbook No. 456, Washington, D.C.: Supt. of Documents, U.S. Government Printing Office.

POMERANZ, Y. P., and MELOAN, C. E. 1978. *Food analysis: theory and practice.* Rev. ed. Westport, Conn.: The AVI Publishing Co.

WATT, B. K., and MERRILL, A. L. 1963. *Composition of foods.* U.S. Dept. of Agr. Handbook No. 8, Washington, D.C.: Supt. of Documents, U.S. Government Printing Office.

Reagents, Materials, and Equipment are listed on page 284.

LABORATORY REPORT

Name_____ Section _____ Date _____

Experiment 22.   DETERMINATION OF FAT BY SOXHLET EXTRACTION

Sample No. _____

	1.	2.
Weight of beaker, thimble, and sample .... _____		_____
Weight of beaker and thimble .... _____		_____
Weight of sample, before extraction .... _____		_____
Weight of extraction flask and fat .... _____		_____
Weight of extraction flask .... _____		_____
Weight of fat .... _____		_____
Percent fat in sample .... _____ %		_____ %

Average percent .... _____ %

QUESTIONS

Name_____ Section _____ Date _____

Experiment 22. DETERMINATION OF FAT BY SOXHLET EXTRACTION

1. a. Write the chemical formula for each of the following:

   Ethyl ether _____ Chloroform _____ Carbon tetrachloride _____

2. List the fat-soluble vitamins, giving BOTH the letter symbol and the name wherever possible.

   _____

   _____

3. Name two phospholipids, other than lecithin, extractable from animal tissue.

   _____

4. Name a specific steroidal compound which is a
   a. Bile salt _____   c. Very common sterol _____

   b. Sex hormone _____   d. Steroidal-type "vitamin,"
                                                 actually a hormone _____

5. From the table below, indicate which food group has the

   a. Highest fat content_____   b. Second highest _____

   c. Lowest fat content (two groups)_____

6. From each of the food groups given in the table below, select the food with the *highest* fat content and indicate the percentage of fat.

   a. Fruit _____   d. Cereal _____

   b. Vegetable _____   e. Nut _____

   c. Meat _____

Proximate Percentage of Fat in Common Foods (Edible Portion)

Fruits		*Vegetables*		*Meats*		*Cereals*		*Nuts*	
apples, not pared	0.6	asparagus	0.2	chicken		whole wheat	2.0	almonds	54
apricots	0.2	lima beans	0.5	broiler	7.2	whole rye	1.7	Brazil	67
bananas	0.2	green beans	0.2	hens	25	oatmeal	7.4	cashew	46
grapefruit	0.1	cabbage	0.2	beef (ground)	21	barley	1.0	peanuts	48
oranges	0.2	corn	1.0	lamb	21	corn flour	2.6	pecans	71
peaches	0.4	carrots	0.2	(composite)				walnuts	64
pears	0.4	peas	0.4	veal chuck	10				
strawberries	0.5	potatoes	0.1	pork	27				
				(composite)					

Note: Fruits, vegetables, and meats are fresh. Meats are medium fat.

Amino Acids and Proteins

General Characteristics of Proteins
Comparison with Lipids and Carbohydrates

INTRODUCTION

Proteins are substances of high molecular weight consisting of alpha amino acids joined together by peptide bonds. All proteins, therefore, contain C, H, O, and N, and most proteins contain sulfur from the amino acids cysteine, cystine, and methionine. In addition, some proteins contain phosphorus, because the hydroxy groups of serine or threonine are esterified with phosphoric acid. A few proteins have mineral elements, such as Fe, Mg, and Cu, as a part of their structures. In contrast to proteins, carbohydrates and fats contain only the elements C, H, and O. There are a few exceptions, however, as in the case of phospholipids and some carbohydrates (hexosamines, chondroitins, etc.).

Heating proteins with soda lime, a fused mixture of calcium oxide and sodium hydroxide, readily degrades the protein and permits the qualitative detection of two elements, N and S, which are not present in the common fats and carbohydrates. Some of the products resulting from the alkaline decomposition of proteins are as follows:

$$\text{Protein} \xrightarrow[\text{heat}]{\text{soda lime}} C + H_2O + N_2 + NH_3 + Na_2S$$

The ammonia nitrogen is easily liberated and detected with indicator paper. The formation of the nonvolatile sodium sulfide (and calcium sulfide) requires stronger and longer heating, and these sulfides must be converted to volatile hydrogen sulfide to confirm the presence of sulfur.

The ignition test readily distinguishes proteins from carbohydrates and fats. Aside from the appearance of carbon, which is general for all of these substances, ignited proteins have a distinctly different odor from fats and carbohydrates. At some time or another, almost everyone has noted the burnt odor of hair or feathers. Again, in contrast to proteins, some carbohydrates and all solid fats readily melt when heated, and, when the temperature is further increased, carbohydrates bubble and lose water as they decompose, while fats can burst into flame. Proteins, on the other hand, merely smoke and stink badly when excessively heated.

Proteins, fats, and carbohydrates differ in solubility characteristics. As noted in the "Determination of Fat" (Expt. 22), proteins and carbohydrates can be separated from fats because only the latter are soluble in nonpolar solvents ($CHCl_3$, ether). The solubility of proteins and carbohydrates in aqueous solutions varies widely. Sugars (mono- and disaccharides) are soluble in water, while certain polysaccharides (cellulose, chitin) and some proteins (hair, feathers) are completely insoluble. Still other polysaccharides (starch, glycogen) and proteins (albumins, globulins) can form colloidal aqueous solutions. The solubility of proteins may be affected by salts and particularly by acids and bases, due to the polar character of certain amino acids. While a protein may be completely soluble in a solution at a low or high pH, it may be insoluble at some intermediate (isoelectric) pH value.

PROCEDURE

A. *Elementary Composition.* In a 150-mm Pyrex test tube place 0.5 g of a protein or proteinaceous material, such as powdered casein, gelatin, egg albumin, wheat gluten, hair, or feathers. Add 0.5 g of soda lime and mix thoroughly with the nickel spatula. Moisten a piece of wide-range Hydrion paper, and then heat the mixture strongly (under hood) while holding the *moist* Hydrion paper in the vapors emerging from the end of the test tube. Record what happens and indicate the pH shown by the Hydrion paper. What *element* is indicated to be present?

Continue to heat the test tube strongly for several minutes; failure to do so will cause the next test to be inconclusive or negative. Allow the tube to cool to room temperature. Acidify the contents of the tube with 5-6 ml of 3 M HCl; the resulting solution *must* be distinctly acid (test with litmus paper). Place a moistened piece of lead acetate paper over the mouth of the test tube and warm the tube *gently* for 10-15 seconds. Describe any changes in the appearance of the paper. Also note any odor at the mouth of the test tube. Which *element* is indicated?

Repeat the entire procedure with a carbohydrate or a fat to determine whether they contain either of the above elements.

NOTE: If laboratory time is limited, the next two parts can be demonstrated by the instructor. Even if some or all of the tests below are omitted, the student should be able to answer all of the questions on the QUESTION sheet.

B. *Ignition Test.* In a 25-mm porcelain crucible place a small amount (enough to cover the bottom of the crucible) of a protein or proteinaceous material (listed in part A), and heat, gently at first and then more strongly, with a flame. Describe in detail whatever changes you observe as the temperature is gradually raised until the bottom of the crucible finally becomes red hot. Did the sample melt, bubble, smoke, catch fire, change color, give off any odor, or leave any unburnable residue? What else did you notice?

Repeat the test with samples of carbohydrate (sucrose, cellulose, or starch) and lipid (tallow or lard). Note particularly any differences between the behavior of fats, proteins, and carbohydrates on ignition.

C. *Solubility.* Test the solubility at room temperature of fresh raw egg white, cottage cheese, wheat gluten, soy protein, and hair or feathers in each of the following solvents: water, 0.1 M NaOH, 0.1 M HCl, and chloroform. Tabulate the results. Carry out whatever laboratory tests you need to do in order to obtain the same information for sucrose, starch, and a fat (solid fat or vegetable oil) and include the results in the table.

Reagents and Materials are listed on page 285.

LABORATORY REPORT

Experiment 23. **GENERAL CHARACTERISTICS OF PROTEINS**
COMPARISON WITH LIPIDS AND CARBOHYDRATES

A. *Elementary Composition. Heating with Soda Lime*

Effect on moist Hydrion paper?_____ pH_____

What *element* is indicated?_____ What *compound* was evolved? _____

Effect on moist lead acetate paper? _____

What *element is* indicated?_____ What *compound* was evolved?_____

Write equations for the following reactions:

Sodium sulfide and dilute HCl_____

Volatile sulfide and lead acetate_____

Results with carbohydrate or lipid. _____

_____

B. *Ignition Tests*

Describe the changes you observed as the temperature was gradually increased until strong ignition (red hot) occurs.

Protein: _____

_____

Carbohydrate: _____

_____

Fat: _____

_____

C. *Solubility*

Make a table showing solubility of the substances tested with the various solvents.

Name _____ Section _____ Date _____

Experiment 23. GENERAL CHARACTERISTICS OF PROTEINS
COMPARISON WITH LIPIDS AND CARBOHYDRATES

1. Name the chemical grouping in a protein structure that contributes to most of the nitrogen found in

   proteins. _____

2. Name the three sulfur-containing amino acids found in most proteins and give the name and structure
   of the group that contains the sulfur.

   Name of amino acid   *Name and structure of sulfur group*

   _____   _____

   _____   _____

   _____   _____

3. Give the chemical formulas of the soda lime components. _____

4. Give the name of a specific conjugated protein

   a. in milk, containing phosphorus.   _____

   b. in blood, containing a metallic element. _____

5. If common carbohydrates contain only C, H, and O, what are the products after ignition in air at dull
   red heat?

$$C,H,O \text{ (carbohydrates, fats)} + O_2 \xrightarrow{\text{heat}}$$

6. If separate samples of proteins, carbohydrates, and fats are heated at a dull red heat until all the black
   carbon is gone, which one(s) would most likely leave a small amount of ash? Explain.

   _____

7. Suppose you had a water-insoluble substance that could be an organic amine, a carboxylic acid, or a
   protein, which substance would be soluble

   a. in dilute acid, but not in dilute alkali? _____

   b. in dilute alkali, but not in dilute acid? _____

   c. in both dilute acid and dilute alkali? _____

8. Name the amino acids whose side chains projecting from the protein "backbone" are most likely to
   have the greatest effect on the ionic character of proteins. List these amino acids according to the nature
   of the ionic charge as indicated below.
   a. Amino acids capable of contributing a + charge:

   _____

   b. Amino acids capable of contributing a − charge:

   _____

9. Which class of substance would you expect to show buffer action—fat, protein, or carbohydrate? Why?

_____

10. a. Give the name of a common *polar* solvent. _____

   nonpolar solvent. _____

 b. In which type of solvent (polar or nonpolar) would the following most likely be soluble?

   Triacylglycerols   _____

   Albumins and globulins _____

   Mono- and disaccharides _____

Color Tests for Proteins and Amino Acids

INTRODUCTION

When the carbonyl group of one amino acid is covalently linked to the α-amino nitrogen of a second amino acid, the resulting compound is called a *dipeptide,* even though it contains only *one* peptide bond. Similarly, the linking of three amino acids yields a tripeptide with *two* peptide bonds; four amino acids form a tetrapeptide with *three* peptide bonds; etc. As more amino acids are joined together, the substance is called a *polypeptide.* As a rule of thumb, a *protein* is a polypeptide chain of at least 40 amino acids with a minimum molecular weight of 5,000. Many proteins contain hundreds of amino acids and have a high molecular weight.

The *biuret test* is commonly used to detect the presence of proteins and peptides by treating a sample with an alkaline solution of dilute copper sulfate to yield a pink-violet to purple-violet color. At least two peptide bonds (tripeptide) are required for a positive test. Biuret (H_2N-CO-NH-CO-NH$_2$), after which the test was named, gives the same color as proteins and peptides. A colored complex that forms between the Cu^{++} and the peptide bonds is the basis for a quantitative estimation of proteins (Expt. 28).

The *ninhydrin reaction* is given by both α-amino acids and proteins. When heated with ninhydrin (triketohydrindene hydrate), the amino acid is oxidized according to the following general equation:

$$\text{Ninhydrin} + \text{RCH(NH}_2)\text{COOH} \longrightarrow \text{RCHO} + \text{NH}_3 + \text{CO}_2 + \text{hydrindantin (reduced ninhydrin)}$$

The liberated NH$_3$ reacts with the hydrindantin (reduced ninhydrin) and a second molecule of ninhydrin to produce a substance with an intense blue to purple color. Proline and hydroxyproline, which lack an α-amino group, yield a yellow color. The free α-amino groups on the N-terminal amino acid of proteins and peptides also liberate NH$_3$ to give the characteristic color with ninhydrin. The ninhydrin reaction is used to locate the α-amino acids as blue to purple spots in paper chromatography (Expt. 25), and permits the quantitative estimation of α-amino acids and peptides in column chromatography.

The yellow color formed in the *xanthoproteic test* is due to the nitration of the aromatic rings in tyrosine and tryptophan by concentrated nitric acid. Under the conditions of the test, phenylalanine does not produce the color because the benzene ring is not activated for nitration.

Some color tests are specific for certain amino acids. The *Millon's test* indicates the presence of tyrosine. When the phenolic group (-⬡-OH) of tyrosine reacts with the reagent (a mixture of Hg^{++}, Hg$_2^{++}$, HNO$_3$ and HNO$_2$), a red color develops.

The *Hopkins-Cole test* for tryptophan results in the formation of a purple ring when a solution containing a mixture of the test sample and Hopkins-Cole reagent is layered over concentrated sulfuric acid. The Hopkins-Cole reagent contains magnesium glyoxylate, and the colored product is formed by the reaction of glyoxylic acid (CHO-COOH) with the indole ring structure of tryptophan in the presence of H$_2$SO$_4$.

The *Sakaguchi test* produces a red color when arginine reacts with α-naphthol and sodium hypobromite (NaOBr). Sakaguchi regarded the color as due to a reaction between the hypobromite and the —NH$_2$ group of the guanidino part (H$_2$N-C-NH-) of arginine.

$$\overset{\displaystyle \|}{\underset{\displaystyle \text{NH}}{}}$$

The *lead acetate test* for labile sulfur in cysteine and cystine produces a brown to black coloration when proteins or peptides containing these amino acids are heated with an alkaline solution of lead acetate resulting in the formation of lead sulfide, PbS. *Folin's test for cystine*, on the other hand, produces its characteristic blue color only with *free* cystine; therefore a protein or peptide must first be hydrolized. Folin's uric acid reagent (a solution of sodium tungstate refluxed with phosphoric acid) has been adapted to the detection as well as the quantitative determination of cystine.

PROCEDURE

A. **General Protein Color Tests.** Perform tests (1), (2), and (3) below on 1-2 ml of 1% solutions of egg albumin, alanine, and sucrose, also 10-20 mg (roughly estimated) of a solid protein suspended in 1-2 ml of water, and on any other substance that may be specified.

On your report sheet indicate the color produced and the chemical grouping of an amino acid or protein that is responsible for the color formation. All tests must be checked by the instructor before discarding them.

(1) *Biuret test.* In separate test tubes thoroughly mix each of the samples to be tested (see above) with 1 ml of 10% NaOH and then add 5-10 drops of 0.1% $CuSO_4$ solution. Compare with a blank tube containing 1-2 ml water and the same amount of NaOH and $CuSO_4$. Describe any color change that occurred.

(2) *Ninhydrin test.* In separate test tubes mix the samples to be tested with 1-2 ml of 0.1% aqueous ninhydrin solution. Also include a tube containing a few drops of dilute ammonium hydroxide. Heat the tubes in a boiling water bath for 3-4 minutes and observe the colors after standing a few minutes. Describe the color changes that occur in each test.

(3) *Xanthoproteic test.* To separate test tubes of the test samples, including one containing 2-3 ml of 0.02% tryptophan solution, add an equal volume of concentrated HNO_3. Heat 1-2 minutes in a boiling water bath. Describe the results in each case.

B. **Color Tests for Specific Amino Acids.** Test the substances that are given in each part below and record your results on the report sheet.

(1) *Millon's test for tyrosine.* In separate test tubes place 2-3 ml of 1% egg albumin, 1% gelatin, 0.02% tyrosine, and 0.02% salicylic acid solutions. Add 3-4 drops of fresh Millon's reagent and heat the tubes in a boiling water bath for 1-2 minutes. Note the color formed. The color may disappear if too much Millon's reagent is used.

(2) *Hopkins-Cole test for tryptophan.* In separate test tubes place 2-3 ml of 1% egg albumin, 1% gelatin, and a few granules of casein suspended in 2-3 ml of water. Add about 3 ml of Hopkins-Cole reagent to each tube and mix thoroughly. Carefully add 5 ml of concentrated H_2SO_4 (wear safety glasses) from the special buret by touching the tip of the stopcock to the inside wall of the test tube and allowing the acid to *slowly* drain down the tube so that the two liquids form separate layers. Hold each tube against a white sheet of paper and observe the color at the zone of contact of the two fluids. If no color appears, swirl the tube gently, but do *not* mix.

(3) *Sakaguchi test for arginine.* To 5 ml of 0.1% cold gelatin solution add 1 ml of 10% NaOH and 1 ml of 0.02% α-naphthol solution. After 3 minutes add 2-4 drops of NaOBr. A strong red color develops but fades quickly. The color can be stabilized by adding urea to destroy the excess hypobromite.

(4) *Lead acetate test for labile sulfur.* To 2-3 ml of 1% egg albumin in a test tube add 5 ml of 5% NaOH and a few crystals of lead acetate. Heat in a boiling water bath for 5-10 minutes, with occasional mixing of the contents of the tube. Describe the color changes.

(5) *Folin's test for free cystine.* To 7 ml of saturated sodium carbonate and 3 ml of 20% sodium sulfite in a test tube add about 1 ml of cystine solution (1 mg/ml). Allow to stand for 5 minutes. Add 1 ml of uric acid reagent. Observe the stability of the color that develops.

Reagents and Materials are listed on page 285.

Name_____ Section _____ Date_____

Experiment 24. COLOR TESTS FOR PROTEINS AND AMINO ACIDS

A. *General Protein Color Tests*

Name of test and formula of reagents	Substance tested	Color produced Test is + or −	Chemical group or amino acid responsible for + test
(1) Biuret test	1% egg albumin		Chemical group:
	1% alanine		
Reagents:	1% sucrose		
	solid protein:		
(2) Ninhydrin test	1% egg albumin		Chemical group:
	1% alanine		
Reagent:	1% sucrose		Reaction with alanine: Show structures.
	solid protein:		
	dilute ammonia		
(3) Xanthoproteic test	1% egg albumin		Amino acids and chemical groups involved:
	1% alanine		
Reagent:	1% sucrose		
	0.02% tryptophan		

B. *Color Tests for Specific Amino Acids*

Name of test and formulas of reagents	Substance tested	Color produced Test is + or −	Amino acid and chemical grouping responsible for + test
(1) Millon's test Reagents:	1% egg albumin		
	1% gelatin		
	0.02% tyrosine		
	0.02% salicylic acid		
(2) Hopkins-Cole test Reagents:	1% egg albumin		
	1% gelatin		
	solid casein		
(3) Sakaguchi test Reagents:	0.1% gelatin		
(4) Lead acetate test Reagents:	1% egg albumin		
(5) Folin's test Reagents:	0.1% cystine		

Name _____ Section _____ Date _____

Experiment 24. COLOR TESTS FOR PROTEINS AND AMINO ACIDS

A. *General Protein Color Tests*

1. Give the name and structure of the grouping that joins the amino acid residues in a protein structure.

   Name _____ Structure _____

2. What type of chemical grouping is the structure described above? _____

3. Will a dipeptide give a positive biuret test? Explain. _____

4. Indicate whether the following substances would give a + or − biuret test and give a reason for your choice.

   a. Tyrosylphenylalanine _____

   b. Insulin _____

   c. Glycylcysteinyltryptophan _____

   d. Alanylglycine _____

5. If the hydrolysis of a protein is continued until the biuret test is negative, what can you conclude about the nature of the products of hydrolysis?

   _____

6. If the average molecular weight of the amino acids in proteins is 125, how many of the chemical groupings named in question 1 would be present in a single protein chain of molecular weight of 75,000? Show work.

7. Write the formulas of alanine and the products obtained from it when it is oxidized by ninhydrin. Encircle the product that reacts with the ninhydrin molecules to form a colored substance.

8. Which amino acids do not produce a purple color with ninhydrin? What color is formed?

   _____

9. Why would the ninhydrin test be more useful and reliable for showing the absence rather than the presence of proteins?

   _____

   _____

10. Do all amino acids give a positive xanthoproteic test? Explain.

   _____

11. Do most proteins give a positive xanthoproteic test? Explain.

_____

B. Color Tests for Specific Amino Acids

1. Give the name and structure of the amino acid responsible for the Millon's test. Encircle the chemical group involved.

   Name_____   Structure:

2. Explain the results of the Millon's test

   a. with gelatin. _____

   b. with salicylic acid. _____

3. Give the name and structure of the amino acid that contains the indole group required for a positive Hopkins-Cole test. Encircle the indole group.

   Name_____   Structure:

4. Name a protein that gives a negative Hopkins-Cole test. _____

5. Give the structure of the two amino acids with labile sulfur and encircle and name the sulfur grouping.

   a. Cysteine structure: Name of group_____

   b. Cystine structure: Name of group_____

6. Why must a protein first be hydrolyzed before applying the Folin's test for cystine?

   _____

7. Write the structure of arginine and encircle the guanidino group.

8. What causes the red color to fade in the Sakaguchi test for arginine, and how could this fading be prevented?

   _____

   _____

Separation and Identification of Amino Acids by Paper Chromatography

INTRODUCTION

The similarity between paper chromatography and thin-layer chromatography, both in principle and in the factors influencing the separation of a mixture into its constituents, has been discussed in Experiment 21. The student should read the introduction to that experiment, noting in particular the remarks on stationary and mobile phases, equilibration of the solvent system, and the distribution coefficient.

The water in the paper fibers acts as the stationary phase as the solvent system flows by capillary action up the paper in "ascending" chromatography, or down the paper in "descending" chromatography. Fine-grain paper gives better resolution of the separated substances than course-grain paper. If paper is cut from larger sheets, it must be done so that the solvent will flow in the same direction as the "grain" of the paper for reproducible results.

Paper chromatography is particularly suitable for separating hydrophilic (water-loving) substances, such as amino acids. In this experiment you will spot the same amino acids on two separate sheets of chromatographic paper, but you will use a different solvent system on each sheet. The two solvent systems have been chosen from the six systems recommended in a publication prepared by the National Academy of Sciences—National Research Council. The one solvent system is acidic, containing acetic acid; the other system is alkaline, containing ammonia. Since the spots of some amino acids may overlap when using one solvent system, they may separate with another system. The student should note in particular how these two different solvent systems affect the R_f values of the basic amino acids (arg, his, lys) and the acidic amino acids (asp, glu).

The overlapping of spots occurs particularly when a large number of components in a mixture are separated. This is the case, for example, when a mixture of 18-20 amino acids in a protein hydrolysate is to be separated. A more complete resolution of such a mixture can be achieved by using two-dimensional chromatography. In this technique the sample is spotted in one corner of a square sheet of paper, and the chromatogram is developed with the first solvent system so that the components of a mixture are partially separated along a vertical line. The paper sheet is dried, rotated 90° from its original position, and then developed with a different solvent system. If the two solvent systems are properly chosen, the separated components will appear, after visualization by the appropriate reagent when necessary, as a fingerprint pattern from which each component hopefully can be identified.

Thin-layer chromatography has many practical advantages compared to paper chromatography (see Experiment 21), which have made TLC increasingly popular. However, some of the advantages of paper chromatography are: (1) paper sheets are easier to prepare and handle than glass plates; (2) it is easier to elute a component from paper than from a glass plate; and (3) the "descending" technique of paper chromatography extends the usefulness of this method over the limited "ascending" technique of TLC. For these reasons, paper chromatography is sometimes the preferred method.

PROCEDURE

A. *Preparation of the Paper.* Prepare two sheets of the 12.5 x 23-cm (5″ x 9″) chromatographic paper. Hold the paper only by the edges; fingerprints will show later when paper is treated with ninhydrin. Draw a light *pencil* line across each paper 12 mm from the bottom edge. Place pencil dots at 2.5-cm intervals on this line, and number the dots from 1 to 7, and the eighth dot mark U for the unknown. At the upper left-hand corner, print your name and the group identification for your seven known amino acids. At the upper right-hand corner of one sheet indicate Solvent A, and on the second sheet, indicate Solvent B.

B. *Spotting of Samples.* The 0.02 M solutions of amino acids are available in dropper bottles. Place two drops of each of your assigned seven amino acids in separately numbered (wax pencil) depressions of your spot plate. Put several drops of distilled water in another depression. (To prevent evaporation of solutions while spotting, it is advisable to put only 3 or 4 amino acid solutions in the spot plate at one time.) To clean the self-filling, 1-μl micropipet, dip its tip into the water and dispense the water on a piece of filter paper. Repeat several times. Each time the spot should be about 3 mm in diameter. Your instructor will demonstrate the proper spotting technique.

Fill the micropipet by touching its tip to amino acid No. 1 of your group and then dispense it on dot No. 1 of one of the chromatographic papers. Refill the micropipet and spot dot No. 1 of the second paper. Rinse the micropipet 3-4 times by alternately filling with water and then dispensing the water on filter paper. Now fill the micropipet with amino acid No. 2 and spot both papers. Repeat the above procedure until all seven amino acids have been spotted on both papers.

The unknown is a mixture of 3 amino acids from your group prepared by your instructor from the same solutions in the dropper bottles that you used for spotting. Therefore, spot your unknown 3 times on each paper to compensate for the dilution of one amino acid by the others in your prepared unknown. Allow the spot to dry momentarily before applying another on top of it. The samples may be spotted during one laboratory period and stored between clean sheets of paper for separation during the next lab period.

C. *Running the Chromatogram.* Prepare aluminum-foil covers on two 800- or 1000-ml beakers, trimming the foil so that it extends about 12 mm down the side of the beaker, thus permitting a full view of the upper part of the chromatogram during development. Place 20 ml of solvent A into one beaker (mark beaker) and 20 ml of solvent B into the second beaker. *Cover beakers immediately.* Make cylinders of the two spotted papers by stapling the ends together. Do not overlap.

Right

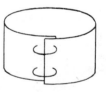

Wrong

Place the paper cylinder marked A in the middle of the beaker containing solvent A with the *spots toward the bottom and cover the beaker immediately* with aluminum foil. Repeat with paper B and beaker B. Never reverse the above procedure by pouring solvent into a beaker containing the paper cylinder, nor jar the beaker so solvent splashes up on the paper. Once the development has started, do not disturb the beaker. Keep the beakers at room temperature; do not set them near a heat source (burner or hot plate).

Allow the solvents to rise about 12 mm (½″) from the top of the paper. Solvent A will require approximately 65-80 minutes; solvent B takes 90-100 minutes. Remove the papers and *immediately* mark the solvent front lightly with a pencil at points about 25 mm from each end and in the middle of the paper. Let the paper cylinders air-dry under your hood. Place them in an oven at 105-110° to complete the solvent removal. Remove the staples.

D. *Treatment with Color Developing Agent.* Grasp the flat paper at the extreme edges and, holding one edge lower than the other, dip it into the ninhydrin reagent solution that will be in a small desiccator or glass pan. Then smoothly raise one hand and lower the other so that the paper is dipped evenly through the reagent. Finally release the trailing edge so that it falls into the dip, then remove the paper, cover the desiccator or glass pan, and dry the paper under the hood at room temperature. (**NOTE:** Acetone is flammable. The acetone solution of ninhydrin must be kept away from flames. Be sure the desiccator or glass pan is covered to prevent the evaporation of the ninhydrin solution.) When dry, place the papers in a 110° oven for 5 minutes. Remove the chromatograms from the oven and draw a pencil line around the edge of each spot.

E. **R_f Calculation.** Draw the best straight line through the points you marked for the solvent front. Put a dot in the center of each encircled spot. Measure the distance of the solvent front from the line on which the samples were spotted (the base line). Measure the distance from the center of the spot to the base line. Calculate the R_f value for each of the amino acids chromatographed according to the equation given below and report the value to *two* decimal places.

$$R_f = \frac{\text{distance spot moved}}{\text{distance solvent moved}}$$

On your report sheet tabulate the distances and the R_f results and make notes also of the exact shade of color produced by each known amino acid. Identify the unknown by comparison with R_f values and colors of the known samples. Sometimes the R_f value of a given amino acid in a mixture may not be exactly the same as its R_f value when it is spotted alone. Therefore, use a little judgment in identifying your unknown. Staple your chromatograms to your laboratory report sheet.

Composition of Solvent Systems.

Solvent A. 1-Butanol, glacial acetic acid, water (volume ratio, 90:10:25.)

Solvent B. 2-Butanol, 3.3% NH_4OH (volume ratio, 75:30).

Amino Acid Group Assignment. You will be assigned one of the groups of seven amino acids. In addition to numbering the base line dots from 1 to 7, write the three-letter abbreviation of the amino acid under the corresponding number. Your unknown will be a mixture of three amino acids, which your instructor will prepare from your assigned group.

Group A	*Group B*	*Group C*
1. arginine	1. glutamic acid	1. aspartic acid
2. aspartic acid	2. glycine	2. proline
3. isoleucine	3. histidine	3. leucine
4. proline	4. leucine	4. lysine
5. serine	5. methionine	5. threonine
6. tryptophan	6. phenylalanine	6. tryptophan
7. valine	7. tyrosine	7. valine
Group D	*Group E*	*Group F*
1. alanine	1. aspartic acid	1. alanine
2. arginine	2. histidine	2. glutamic acid
3. glutamic acid	3. isoleucine	3. isoleucine
4. phenylalanine	4. methionine	4. proline
5. leucine	5. proline	5. lysine
6. tryptophan	6. serine	6. tryptophan
7. valine	7. tyrosine	7. valine

REFERENCES

National Academy of Sciences—National Research Council. 1967. *Specifications and criteria for biochemical compounds.* 2d ed. Washington, D.C.: National Academy of Sciences, pp. 4-5.

UMBREIT, M. W.; BURRIS, R. H.; and STAUFFER, H. H. 1972. *Manometric and biochemical techniques.* 5th ed. Minneapolis, Minn.: Burgess Publishing Co., chap. 15.

Reagents and Materials are listed on page 287.

Name _____ Section _____ Date _____

Experiment 25.   SEPARATION AND IDENTIFICATION OF AMINO ACIDS BY PAPER CHROMATOGRAPHY

Amino Acid Group _____

	—Solvent system A—		*—Solvent system B—*	
Name of amino acid	*mm from base line*	*R_f value*	*mm from base line*	*R_f value*
(solvent front)	_____	1.00	_____	1.00
1. _____	_____	_____	_____	_____
2. _____	_____	_____	_____	_____
3. _____	_____	_____	_____	_____
4. _____	_____	_____	_____	_____
5. _____	_____	_____	_____	_____
6. _____	_____	_____	_____	_____
7. _____	_____	_____	_____	_____

Unknown Sample No. _____

Unknown amino acids:

_____	_____	_____	_____	_____
_____	_____	_____	_____	_____
_____	_____	_____	_____	_____

Components of Solvent A: _____

Components of Solvent B: _____

Name _____ Section _____ Date _____

Experiment 25.   SEPARATION AND IDENTIFICATION OF AMINO ACIDS BY PAPER CHROMATOGRAPHY

1. What is the stationary phase in this experiment? _____

2. What is meant by the "mobile phase"? _____

   _____

3. Is "ascending" or "descending" chromatography used in this experiment? Explain.

   _____

4. What type of substances are more suitably separated by paper chromatography than by thin-layer chromatography?

   _____

5. Make the following conversions:

   a.   1 μl = _____ liter   e. 1 $\times$ 10^{-5} liters = _____ μl

   b. 150 μl = _____ liter   f.   0.025 ml   = _____ μl

   c.   1 μl = _____ ml   g.   25 μg   = _____ g

   d. 250 μl = _____ ml   h.   200 μg   = _____ mg

6. If 1 μl of 0.02 M alanine (M.W. 89) is spotted, how many micrograms (μg) of alanine has been applied? How many milligrams? Show work.

7. Which amino acid, present in some of the assigned groups, has a yellow color with ninhydrin?

   _____

8. What is the purpose of using two different solvent systems in this experiment?

   _____

   _____

9. Solvent A is acidic (acetic acid) and Solvent B is alkaline (ammonia).
   a. What is the structure and predominant net charge of lysine in each solvent system?

   Solvent A   *Solvent B*

   Lysine structure: Lysine structure:

   Net charge _____   Net charge _____

b. Is lysine more polar in solvent A or B? Explain.

_____

c. Will lysine have a higher R_f value in solvent A or B? Explain.

_____

(Note: The above reasoning applies with histidine and arginine. Check the R_f values of the basic amino acid in your group to see if they agree with (c) above.)

10. Again, if solvent A is acidic and B is alkaline
   a. What is the structure and predominant net charge of aspartic acid in each solvent system?

Solvent A	_Solvent B_
Aspartic acid structure:	Aspartic acid structure:
Net charge _____	Net charge _____

   b. Is aspartic acid more polar in solvent A or B? Explain.

_____

   c. Will aspartic acid have a higher R_f value in solvent A or B? Explain.

_____

(Note: Glutamic acid acts similarly. Check the R_f values of the acidic amino acid in your group to see if they agree with (c) above.)

11. How would you modify the procedure in this experiment so that the separation of amino acids could be done by two-dimensional chromatography?

Precipitation of Proteins

INTRODUCTION

When the secondary and tertiary structures of native proteins are altered by chemical or physical means, the protein molecules tend to agglomerate and precipitate, and the protein becomes *denatured*. Depending upon conditions, this denaturation can be either irreversible (excessive heating or extreme pH changes) or reversible (treatment with ammonium sulfate). Denaturation of a protein is accompanied by loss of its biological activity. In general, protein precipitation is due to a disruption of hydrogen bonds, ionic (saltlike) bonds, and sometimes the stronger, covalent disulfide bonds, all of which hold together the protein structure.

As amphiprotic substances, proteins can accept or give up protons. The amino groups, and certain nitrogen atoms of the basic amino acid residues (lysine, arginine, and histidine), and the carboxyl groups of the acidic amino acids (aspartic and glutamic) can be converted to positive and negative charged groups, respectively, depending on the pH of the medium. When *strong* acids are added to a neutral protein solution, the carboxylate groups become undissociated carboxyl groups, and the nitrogen atoms with available electrons become protonated resulting in a positively charged protein molecule. Because the ionic (salt) bridges and the hydrogen bonds are disrupted, the protein precipitates.

When *heavy metal cations* are added to protein solutions, the metal ions combine with the negatively charged groups to form an insoluble metal ion proteinate. Certain *acidic* reagents (trichloroacetic acid, tannic acid, phosphotungstic acid) combine with proteins to form insoluble protein salts (e.g., protein tannate). These acids are called *alkaloidal reagents* because they combine with alkaloids (organic bases from plants, e.g., morphine, strychnine, cocaine) to form precipitates. Phosphotungstic acid and trichloroacetic acid are often used to prepare protein-free solutions for the analysis of small molecules as, for example, glucose and amino acids.

Proteins are frequently precipitated with concentrated or saturated solutions of salts, such as ammonium sulfate or magnesium sulfate. The solubility of a protein in a salt solution is a function of the "ionic strength" of the solution, where the ionic strength depends not only on the concentration of the cations and anions but also on the electrical charge (valence) of each ion. The solubility of proteins decreases as the salt concentration (ionic strength) increases, and the protein eventually precipitates completely, a process that is called *salting-out*. One of the many factors involved in the salting-out of proteins is the hydration of the inorganic ions, which disrupts the water of hydration surrounding and protecting the protein molecules, causing them to coagulate. The proteins in a mixture may differ in their solubility at a given salt concentration and may precipitate at different levels of ionic strength. For example, the albumins and globulins in egg white can be separated if the egg white solution is only half-saturated with respect to ammonium sulfate; the globulins will precipitate, but the albumins will not. After filtration, the albumins can be precipitated from the filtrate by further addition of ammonium sulfate to the saturation point. In this experiment the two proteins will be jointly precipitated, and the presence or absence of the proteins will be confirmed with the biuret test. Proteins denatured by salts can often be redissolved (renatured) without loss of biological activity, e.g., the isolation and purification of enzymes.

Proteins can also be precipitated from aqueous solutions by the addition of polar organic solvents, such as ethanol and acetone. These solvents interfere with hydrogen bondings within the protein molecule, as well as the hydrogen bonding between water and protein molecules. In addition, electrical charges on the ionizable groups on the side chains of proteins are affected. All these factors permit the aggregation and precipitation of proteins.

Physical means, such as vigorous shaking or whipping, will also coagulate proteins. The *Sevag test* illustrates the denaturation of a protein by physical shaking. The immiscible chloroform used in the test assists in the agitation of the protein and enables one to observe the insoluble precipitate that forms.

PROCEDURE

Show all tests and your recorded observations to your instructor.

A. *Strong Acids.* In each of three test tubes place 3 ml of clear 1% egg white solution. To one add 1-2 ml of concentrated HCl, to another HNO_3, and to the third H_2SO_4. Note and describe the changes that occur.

B. *Heavy Metal Cations.* To 2-3 ml of 1% egg white solution add 0.2 M $CuSO_4$ solution drop by drop. Repeat with 0.2M lead acetate and mercuric chloride solutions. Also add one of the heavy metal salt solutions to 2-3 ml of 1% urea.

C. *Alkaloidal Reagents.* To 2-3 ml of 1% egg white solution add 10% trichloroacetic acid drop by drop, mix well, and describe the result. Repeat using 5% aqueous tannic acid and 20% phosphotungstic acid solutions. Also carry out the test by adding one of the acids to 1% solutions of urea and of alanine.

D. *Salting-Out.* To 20 ml of 1% egg white solution in a small beaker add 10 g of solid ammonium sulfate. Dissolve the ammonium sulfate by warming the mixture *gently* with continuous stirring. Do *not* exceed 40°C (warm to the touch) or you will coagulate the egg white into a curd. The egg proteins will give the mixture a milky appearance as they are salted-out of solution, and it will be difficult to ascertain when all of the $(NH_4)_2SO_4$ has dissolved. After the salt has dissolved, filter the mixture through a medium-grade filter paper (e.g., Whatman No. 40) collecting the filtrate in a clean, small beaker.

Test 2 ml of the clear filtrate for protein with the biuret test (Expt. 24), using about 1.5 ml of the NaOH instead of the 1 ml specified. This slight excess of NaOH destroys the ammonium ion, which intensifies the interfering blue color of the Cu^{++} added. A blue color is not a positive test for protein.

Remove a small portion of the precipitate with the tip of your spatula and dissolve it in 2-3 ml of water in a test tube. Repeat the biuret test and compare the results with the test on the filtrate.

E. *Organic Solvents.* To 2-3 ml of 1% egg white solution add an equal volume of 95% ethanol. Mix thoroughly and note the results. Repeat the test with acetone.

F. *Sevag Test.* To 2-3 ml of 1% egg white solution add 1 ml of chloroform and shake the test tube *vigorously* for a minute. Allow to stand and look for indications of a precipitate at the boundary between the two liquids.

Reagents and Materials are listed on page 288.

LABORATORY REPORT

Name______________________ Section______________ Date______________

Experiment 26. PRECIPITATION OF PROTEINS

A. *Strong Acids*

Describe the changes that occur when each of the strong acids is added to 1% egg white solution.

HCl ______________________

HNO_3 ______________________

H_2SO_4 ______________________

B. *Heavy Metal Cations*

Describe the changes that occur when each of the heavy metal cations is added to 1% egg white solution.

Cu^{++} ______________________

Pb^{++} ______________________

Hg^{++} ______________________

What happens when one of the heavy metal cations is added to 1% urea?

____

C. *Alkaloidal Reagents*

Describe what occurs when each of the alkaloidal reagents is added to egg white solutions.

Trichloroacetic acid ______________________

Tannic acid ______________________

Phosphotungstic acid ______________________

What happens when one of the alkaloidal solutions is added to

a solution of urea? ______________________

a solution of alanine? ______________________

D. *Salting-Out*

Results of biuret test

on filtrate ______________ on precipitate ______________

Conclusions ______________________

____

E. *Organic Solvents*

What happens when ethanol or acetone is added to egg white solution?

_____

F. *Sevag Test*

Results after vigorously shaking egg white solution with $CHCl_3$.

_____

Name_____ Section_____ Date_____

Experiment 26. PRECIPITATION OF PROTEINS

1. What is meant by the following types of protein structures?

    a. Secondary structure. _____

    _____

    b. Tertiary structure. _____

    _____

2. a. Name the types of bonds and linkages that maintain the secondary and tertiary structures of pro-

    teins. _____

    _____

    b. Which type of linkage or bonding is most resistant to denaturation?_____

3. What is meant by denaturation of a protein?

    _____

    _____

4. How do the results of adding heavy metal ions and alkaloidal reagents to protein differ from the re-
    sults obtained by adding these reagents to urea or amino acids?

    _____

5. What type of chemical group is present in both urea and proteins?_____

6. Why are urea and amino acids chosen in this experiment as representatives of low molecular weight,

    nonprotein-nitrogen substances? _____

    _____

7. How could you separate proteins from these nonprotein substances?

    _____

    _____

8. Why is egg white used as an antidote in lead or mercury poisoning? Why must the stomach be pumped
    immediately after use of egg white?

    _____

    _____

9. Give the net charge (+ or −) on the proteins in egg white, and list the amino acids whose side groups are mainly responsible for this charge

   a. in an acid solution (pH 2-3).   _____

   b. in an alkaline solution (pH 10-13). _____

10. What groups on the protein molecule are responsible for precipitation

    by heavy metal cations?_____

    by alkaloidal reagents? _____

11. At what pH is a protein least soluble? Why?

    _____

    _____

12. Explain the effect of the high concentration of ammonium sulfate on the solubility of proteins.

    _____

    _____

13. Was the denaturation of egg-white proteins by ammonium sulfate reversible or irreversible? Explain on the basis of your observations in this experiment.

    _____

    _____

14. In what way does the addition of alcohols and acetone affect the secondary and tertiary structures of proteins? Are all proteins affected the same with a given volume of polar organic solvent?

    _____

    _____

15. Give an example of the denaturation of egg white by whipping as a practical application.

    _____

Kjeldahl Semimicrodetermination of Nitrogen and Crude Protein

INTRODUCTION

Analysis on a semimicro scale is convenient when homogeneous material is available. When relatively heterogeneous substances are analyzed, as in the determination of crude protein in many foods and brewing materials, the Kjeldahl macromethod is recommended. In either case, the protein and other organic matter is digested (oxidized) and the nitrogen is reduced to ammonium sulfate.

$$\text{Protein and other organic matter} \xrightarrow[\text{HgSO}_4,\text{ heat}]{\text{conc. H}_2\text{SO}_4,\text{ K}_2\text{SO}_4} (NH_4)_2SO_4 + CO_2 + H_2O$$

The K_2SO_4 is added to increase the boiling point of H_2SO_4 and make the oxidation-reduction more effective. Mercuric sulfate serves as a catalyst to further reduce the digestion time. Hiller, Plazin, and Van Slyke have investigated Kjeldahl methods carefully and have concluded that mercuric ion is superior to other catalysts recommended for the digestion.

After the digestion is complete, the mixture is diluted with distilled water. A solution containing sodium hydroxide and sodium thiosulfate ($Na_2S_2O_3$) is added to make the solution alkaline, and thereby convert the ammonium ion to volatile ammonia. The thiosulfate reduces the mercuric ion and prevents the formation of a complex between Hg^{++} and NH_3 so that the distillation of ammonia from the alkaline solution may be complete. The condensate from the distillation is received below the surface of a boric acid solution to prevent the loss of ammonia during distillation and prior to titration. This is a direct titration of the NH_3 captured by the boric acid; the boric acid is a sufficiently strong acid to catch the NH_3 but is weak enough so that the NH_3 may be titrated in its presence.

The percentage of crude protein in a substance is obtained by determining the percentage of nitrogen and multiplying this quantity by the factor 6.25. The basis for this calculation is the assumption that protein contains 16 percent nitrogen; hence the percentage of nitrogen multiplied by the factor 6.25 gives the percentage of protein. This method of calculation is not strictly correct, as some proteins contain more and some less than 16 percent of nitrogen, but for general purposes 6.25 is used. A second assumption is made in this calculation, viz., that all the nitrogen in the substance is present in the form of protein, whereas more or less nitrogen is always present as amides, amino acids, or other nitrogen compounds. For this reason the figures in tables giving the protein contents of foodstuffs are likely to be high.

The semimicro method is useful for samples containing 0.6 to 25 mg nitrogen if the concentration of the acid used in the titration is between 0.02 and 0.05 N. In this experiment a series of dried blood samples is used to illustrate the procedure, although any other homogeneous material may be used.

PROCEDURE

A. *Digestion of Protein.* Accurately weigh 140-160 mg of dried blood sample (or other material) on a cigarette paper using the analytical balance. Carefully roll the cigarette paper into a ball and drop it into a dry 100-ml Kjeldahl flask. Run duplicate samples and a blank. The blank flask will contain the cigarette paper and the reagents, but no sample. (**NOTE:** To save on equipment space and time, the instructor or a few designated students may determine the reagent blank.) On the frosted area on the flask, or on a piece of white tape placed at this location, identify the weighed samples as 1 and 2, and write your initials. To each flask add 1.5 g (¼ level teaspoon) of K_2SO_4, 1.5 ml of mercuric sulfate solution, 3 ml of concentrated H_2SO_4, and 2 glass beads (prevents bumping during later distillation).

Heat the flasks on the digestion rack to remove the water and then increase the heat so that the solution boils constantly with slight motion. Occasionally swirl the flask (use the test tube holder) to remove the black carbonaceous material adhering to the upper sides of the flask. After complete clearing, continue to boil gently for 30 minutes. Remove the flasks from the rack and set them to cool in a large beaker under the hood for 5 to 10 minutes. Do not allow the solution to solidify in the flask.

Carefully add 50 ml of water to the Kjeldahl flask, slowly with gentle swirling. Do not point the flask at anyone and be sure you are wearing your safety goggles. If time does not permit, stopper the Kjeldahl flask and distill during the next lab period.

NOTE: If the digestion mixture solidifies due to longer standing, particularly from one laboratory period to the next, then the 50 ml of distilled water is to be added as follows: Add only 5 ml of water to the digestion mixture and swirl. The interaction of the water and the acid will generate heat, which will aid in dissolving the solid matter. Add a second 5-ml portion of water and mix. Repeat a third and fourth time and then add the remainder of the water so that a total volume of 50 ml has been added.

B. *Distillation of Ammonia.* Add 10 ml of 2% boric acid solution (use a graduate) to each of two marked 125-ml Erlenmeyer flasks. Wash off the glass tubing below the condenser with a stream of distilled water. Set the flasks underneath the condenser of the distillation unit so that the glass delivery tube is *below* the surface of the boric acid.

The instructor will demonstrate the proper technique for the distillation procedure given below. Follow each step of the procedure carefully. Be sure that the numbers on the Kjeldahl flasks match those on the Erlenmeyer flasks. *Wear safety goggles!*

Without shaking the *cool* Kjeldahl flask, add 10 ml (graduate) of 13 N NaOH (containing 5% $Na_2S_2O_3$) by flowing it down the wall of the inclined Kjeldahl flask; it will layer at the bottom of the flask. Immediately attach the flask to the distillation set-up and turn on the heater. *Now swirl the flask to mix the contents.* Continue the distillation until about half of the contents of the flask has distilled over into the boric acid. Lower the platform holding the Erlenmeyer flask so that the delivery tube is out of the solution. Turn off the heater, but continue the distillation for a minute or two, making use of the residual heat. Now rinse the outside of the delivery tube with a stream of distilled water and remove the Erlenmeyer flask.

NOTE: It is important to mix the contents of the Kjeldahl flask after the flask is attached to the rubber stopper and the heater is turned on. If thorough mixing is not done, the lower alkaline layer is heated to the boiling point and reacts violently with the acid layer above. As a result the stopper may blow off or the alkaline solution may boil over into the boric acid solution. This means you will have to clean out the apparatus by distilling water through it to get rid of the contaminating alkali. A new sample must be digested and distilled.

C. *Titration of Ammonia.* Titrate the ammonia in the boric acid solution with standard acid (about 0.05 N). The weak boric acid does not interfere with the titration. Use two drops of Tashiro's indicator (methylene blue and methyl red). This indicator turns from green through gray to purple, and the gray serves as a satisfactory end point.

Calculations of Percent Nitrogen and Protein. The milliequivalents of H_2SO_4 used in the titration are equal to the milliequivalents of ammonia titrated which in turn equals the milliequivalents of nitrogen in the ammonia derived during digestion of the original sample.

REFERENCES

Aminco Report No. 104. 1959. The determination of nitrogen by the Kjeldahl procedure including digestion, distillation, and titration. Silver Spring, Md.: American Instrument Co., Inc.

POMERANZ, Y. P., and MELOAN, C. E. 1978. *Food analysis: theory and practice.* Rev. ed. Westport, Conn.: The AVI Publishing Co.

Reagents, Materials, and Equipment are listed on page 289.

LABORATORY REPORT

Name_____ Section_____ Date_____

Experiment 27. KJELDAHL SEMIMICRODETERMINATION OF NITROGEN AND CRUDE PROTEIN

Sample No. _____

	1.	2.
Weight of paper and sample	_____	_____
Weight of paper	_____	_____
Weight of sample	_____	_____
Ml of 0.05 N H_2SO_4 (sample)	_____	_____
Ml of 0.05 N H_2SO_4 (blank)	_____	_____
Net ml H_2SO_4 (sample)	_____	_____
Meq of N = Meq H_2SO_4	_____	_____
Meq wt of N (in mg)	_____	_____
Milligrams N (meq N $\times$ meq wt N)	_____	_____
Percent nitrogen (mg N $\div$ mg sample) $\times$ 100	_____	_____
Percent crude protein (%N $\times$ 6.25)	_____	_____

Name_____ Section_____ Date_____

Experiment 27.   KJELDAHL SEMIMICRODETERMINATION OF NITROGEN AND CRUDE PROTEIN

1. Give the purpose of each reagent in the digestion and distillation procedures.
   a. Digestion of protein

   Conc. H_2SO_4 _____

   K_2SO_4 _____

   $HgSO_4$ _____

   b. Distillation

   13 N NaOH _____

   Sodium thiosulfate _____

   2% Boric acid _____

2. Write equations for the chemical reactions involved in each step.

   a. Digestion of protein _____

   b. Liberation of ammonia by alkali _____

   c. Titration of ammonia _____

3. Give two reasons why the factor, 6.25, may not always be correct for the calculation of percent crude protein.

   a._____

   b._____

4. Which amino acids, if present in large amounts, would give high results when the factor 6.25 is used for the conversion of percent nitrogen to percent protein? Indicate the number of N atoms in each amino acid.

   _____

   _____

5. Name a protein associated with nucleic acids that contains a large amount of the amino acids listed in (4) above.

   _____

Biuret Determination of Protein

INTRODUCTION

When a protein is treated with an alkaline solution of dilute copper sulfate, the Cu^{++} ions form a pink to violet coordination complex with the nitrogen atoms of the peptide bonds (see *biuret reaction*, Expt. 24). Since the color intensity is reasonably reproducible with a given protein, and the copper complexes with different proteins have similar absorption spectra, the biuret reaction is a convenient and rapid method for the quantitative determination of protein. Although serum albumin or lysozyme can be used as a standard in this procedure, they may not always be the best standards for the particular protein being analyzed.

The protein under test must give a clear solution at a concentration of 1 mg per ml, or greater, in 0.6 M NaOH, which is the approximate alkalinity of the mixture of the sample and biuret reagent. Under these conditions, the protein itself should have practically no absorption at the wavelength (540 nm) at which the color intensity of the complex is measured. The protein solution should not contain tris (hydroxymethyl)-aminomethane, free histidine, or an excess of ammonium ions.

Another colorimetric procedure for soluble protein is the Lowry method, which uses the Folin-Ciocalteu phenol reagent. While the biuret method is very useful in the range of 1 to 20 mg of protein, the Lowry method is sensitive in the lower range of 5 to 300 micrograms of protein per ml. In the Lowry method, two color reactions occur with the protein. First, a low-level biuret color is produced when the alkaline copper reagent reacts with the peptide bonds of the small amount of protein in the reaction mixture; then a bluish-green color is formed when the Folin-Ciocalteu phenol reagent (a mixture of phosphomolybdate and phosphotungstate) reacts with tyrosine and tryptophan residues in the protein molecule. Generally the tyrosine and tryptophan content of most proteins is sufficiently uniform to make the Lowry assay for proteins commonly acceptable, provided an appropriate pure protein is selected as a reference standard. For example, bovine serum albumin, which is satisfactory for a protein standard in the biuret method, is not suitable in the Lowry procedure because of its low tyrosine and tryptophan content. In contrast, egg white lysozyme can be used as a protein standard in either of the two colorimetric methods. The Lowry method is particularly useful in protein purification procedures (including enzymes) where small amounts of protein samples are more practical and less wasteful.

PROCEDURE

Preparation and Comparison of Standards and Unknowns. Set up a series of test tubes for a blank, the protein standards, and the unknown as indicated in the table below. For an "unknown," the instructor will give the student 4-5 ml of serum albumin or lysozyme solution at a concentration of 1-10 mg per ml in 1% NaCl. Set up the "unknown" in duplicate.

		Protein standards				Unknown	
Tube No.	*Blank*	*1*	*2*	*3*	*4*	*5*	*6*
Standard protein. 10 mg/ml	0.00	0.20	0.40	0.60	0.80	1.0 ml unknown	
1% NaCl solution	1.00	0.80	0.60	0.40	0.20	in each tube	
Biuret reagent	4.0	4.0	4.0	4.0	4.0	4.0	4.0

Add the biuret reagent last, then thoroughly mix the contents of each tube. Let stand for 30 minutes at room temperature. Read the absorbance of the standards and unknown against the reagent blank. (See page 21 for details on use of the colorimeter or spectrophotometer.)

Plot the absorbance readings (vertical axis) against the concentrations of standard protein expressed as mg per tube (horizontal axis). Determine the mg/ml protein of your unknown.

REFERENCES

GORNALL, A. G.; BARDAWILL, C. J.; and DAVID, M. M. 1949. Determination of serum proteins by means of the biuret reagent. *J. Biol. Chem.* 177:751.

LOWRY, O. H.; ROSEBROUGH, N. J.; FARR, A. L.; and RANDALL, R. J. 1951. Protein measurement with the Folin phenol reagent. *J. Biol. Chem.* 193:265-75.

Reagents, Materials, and Equipment are listed on page 289.

Name _____ Section _____ Date _____

Experiment 28. BIURET DETERMINATION OF PROTEIN

Data and results

Complete the table and prepare a standard curve by plotting absorbance (vertical axis) against serum albumin or lysozyme concentration as mg per tube. Determine the protein concentration of your unknown from this graph.

	Protein standards				Unknown	
Tube No.	1	2	3	4	5	6
Absorbance						
Protein mg/tube						

Unknown No. _____ Name of standard protein _____ Unknown _____ mg/ml

QUESTIONS

Name_____Section_____Date_____

Experiment 28. BIURET DETERMINATION OF PROTEIN

1. Why will ammonium ions and Tris buffer interfere if present in sufficient amount?

_____

2. Why does the biuret reagent contain sodium potassium tartrate?

_____

3. What is the purpose of the "blank" tube? Why not read absorbancies of samples against water?

_____

_____

4. What are some of the advantages of the biuret method? Disadvantages?

Advantages _____

Disadvantages _____

5. What part of the protein structure produces a color with the reagents used

a. in the biuret method?_____

b. in the Lowry method?_____

6. Why is bovine serum albumin unsuitable as a protein standard in the Lowry (Folin-Ciocalteu) method?

_____

7. Compare the range of protein concentration used in the Lowry method with that of the biuret method in terms of mg/ml.

Lowry method_____ Biuret method _____

8. List the reagents used in preparing the biuret reagent.

_____

_____

Electrophoresis of Serum Proteins

INTRODUCTION

Electrophoresis is the movement of charged molecules or ions to one electrode under the influence of an electrical potential. Many of the amino acid side chains of proteins contain ionizable groups that cause proteins in solution to act as charged polyelectrolytes that can migrate in an electrical field. If the proteins in a mixture have different isoelectric points, the net charge on each protein will differ sufficiently to permit their separation. Generally an alkaline buffer is used so that the components of a protein mixture acquire a negative net charge and will move toward the anode when a potential is applied. Some of the factors affecting the rate of migration of proteins are (1) the magnitude of the net charge on the protein at a given buffer pH; (2) the size and shape of the protein moleclues; (3) the ionic strength of the buffer (i.e., buffer concentration); and (4) the applied electrical potential.

The separation of serum proteins in this experiment on a cellulose polyacetate strip is referred to as the standard zone electrophoresis. In this technique the sample is applied as a narrow zone or band on the support strip whose ends dip into the buffer solution contained in the outer cathode and anode compartments of the electrophoresis chamber. Platinum wires make contact with the buffer and the power supply. When a current passes through the buffer and the support strip, the proteins in the sample separate as they move toward the anode. After electrophoresis the protein bands are stained with a dye and the excess dye is removed to permit permanent mounting of the cleared, transparent strip or evaluation by optical scanning with a densitometer. A quantitative colorimetric analysis of the protein fractions can be made by cutting the separated zones from the strip and eluting the protein-bound dye with sodium hydroxide.

The cellulose polyacetate strip offers little resistance to the flow of current and is inert to the proteins and to the dye used. Compared to paper and other support media, cellulose acetate strips cut the running time from 16-24 hours to 30-75 minutes. Because the cellulose polyacetate does not absorb proteins, clear distinct bands with no tailing between them are produced. Cellulose polyacetate is free of contaminants that might interfere with the elution of the protein fractions or the isolation of enzymes that could be altered or destroyed by impurities in a support medium. Cellulose polyactetate strips are a more versatile support medium than paper. For example, hemoglobins, LDH isoenzymes, and lipoproteins are more difficult or impossible to separate on paper.

Electrophoresis plays an important role in clinical diagnosis. Most body fluids can be separated into well-defined components by electrophoresis. Certain pathological conditions cause pronounced changes that can be detected by the absence, the position, or the amounts of these components after electrophoresis. For example, the separation of serum proteins into albumin and globulin fractions, as done in this experiment, has important diagnostic significance. Variations in protein fractions from their normal values occur in macroglobulinemia and multiple myeloma. The normal albumin/globulin ratio is 1.1/1 to 2.0/1, while in the abnormal conditions mentioned above, the A/G ratio is less than 1 (e.g., 0.25/1 to 0.4/1 in macroglobulinemia, and 0.65/1 to 0.85/1 in multiple myeloma).

PROCEDURE

A. *Preparation of Electrophoresis Strips and Chamber.* Students are to work in pairs to assist each other. A student will need one electrophoresis strip if only the cleared strip is prepared (part E), but a second strip is needed if a quantitative elution is also performed (part F).

To obtain the best results, it is essential to keep the cellulose polyacetate strips as wet as possible at all times. Faster evaporation along the edges of drier strips causes bowing of the fractions. The serum sample diffuses readily into a wet cellulose acetate strip, and sample application is, therefore, much more uniform.

Soak the strips in fresh buffer until wet. *Do not omit this step.* Soaking produces a colloidal change that brings the cellulose polyacetate back to its original gel structure. It is important that each strip be floated individually on the buffer and be allowed to wet thoroughly before immersion. If the strip is immersed without complete wetting, air pockets will be trapped in the membrane, and the cellulose acetate will not be activated to support electrophoresis in these areas. The instructor will soak a sufficient quantity of strips for the class.

Do *not* connect the power supply to the chamber. This will be done *by the instructor* after all strips have been placed in the chamber.

The pH 8.8 buffer should be refrigerated until ready to use. Cold buffer improves the electrophoresis of the proteins. Fill the chamber with 900 ml of the buffer. The older, smaller chambers require only 500 ml of buffer. Level the buffer by tilting the chamber so that the fluid just runs over the center barrier and equalizes in the two compartments. The chamber should be kept covered and the cover should be removed temporarily only when placing strips in the chamber.

The electrophoresis chamber should be arranged on the laboratory bench with the *cathode* compartment to the *left*. When using the Gelman deluxe chamber, the electrodes should point *away* from you and the polarity toggle switch should point to the *left* (cathode). When using the older, smaller chamber, the electrode terminals should point *toward* you, with the black terminal (cathode) to the *left*. It is important that the applied sample on the electrophoresis strip always be closer to the cathode than to the anode.

B. *Application of the Serum Sample.* Your instructor will demonstrate the filling of the applicator and the transfer of the serum to the electrophoresis strip. Uniform filling and application of the sample is very important to prevent the distortion of the protein during electrophoresis.

Insert a long, 10-μl capillary tube into a micropipet holder that is used for spotting TLC and paper chromatograms (Expt. 21 and 25). Place the tip of the capillary tube into about 1 ml of serum contained in a small test tube and allow the serum to rise one-half to two-thirds the length of the tube. To accomplish this, hold the test tube and the capillary tube as close to the horizontal as possible without spilling serum from the test tube.

Touch the tip of the capillary tube to the parallel wires of the applicator and allow the sample to drain and adhere to the wires as the capillary tip is moved back and forth along the wires. If the capilliary pipet fails to self-drain, apply a gentle pressure to the rubber bulb to start the drainage of the serum. Be sure that a *continuous film* of serum is applied without gaps, otherwise discontinuous bands will result later. If the wires accidentally become overloaded, excess sample can be removed by holding the capillary tube horizontally while touching it to the wires. If a poor application has been made, the wires can be wiped clean with a tissue and the application process repeated. (**NOTE:** *Never pull on the wires* when cleaning them. Many expensive applicators have been ruined because one or both wires were pulled loose.) A properly filled applicator contains 1.6 μl of serum sample between the 18-mm (¾") length of the parallel wires

When the first student has filled the applicator properly, the second student will take an electrophoresis strip from the soaking tray with forceps and place it on an absorbent pad that has been previously marked with a pencil line 5 cm from the left end and has been moistened with buffer solution. The end of the strip is aligned with the left edge of the pad. Blot the strip *very gently* with another absorbent pad. *Only excess surface buffer should be removed.* Keep the strips as wet as possible to ensure proper diffusion of the sample into the strip. After blotting, leave the strip on the moist absorbent pad to provide a resilient background for sample application.

Apply the sample 5 cm from the left end of the strip using the pencil line on the blotter pad as a guide. Straddle the applicator over the electrophoresis strip, positioning it so that the parallel wires containing the sample are in line with the pencil mark. Press down the button of the applicator so that the wires are held firmly against the strip for a few seconds. The sample is transferred as a straight line of uniform density.

Only *one* application of sample is made on the strip if the strip is to be cleared after electrophoresis in accordance with part E. *Write your initials* with a ball-point marking pen about 2.5 cm (1″) from the right end of the strip.

NOTE: *If the strip is to be used for quantitative elution, refill the applicator and apply two additional portions on exactly the same place as the first application.* The 5 μl of sample from this triple application will give better quantitative results in part F.

Lift the strip from the pad, holding it horizontally taut, and carefully place it across the electrophoresis chamber dividers so that the ends of the strip dip into the buffer solution in the two outer compartments. The point of sample application must be *about 1 cm from the outer cathode support bridge of the chamber,* otherwise the sample will move in the wrong direction (into the buffer solution) when the voltage is applied.

Tension the strip to the support bridges by pressing the strip to the bridge and smoothing the ends of the strip on the bridge. Do not allow buffer to splash or drip on the strip. The strips are held in position by arrow-shaped Magna-Grips. Place the Magna-Grips over the strip on the top of the support bridges with the arrows pointing toward the end of the chamber where the electrodes are located. Slowly move the Magna-Grips down the side of each bridge until they hold firmly. The top edge of the Magna-Grips should be approximately 6 mm below the top of the support bridge. There must be no sag in the strip. *Replace the cover on the chamber as soon as the strip is tensioned to prevent evaporation from the strip. Keep the chamber covered* while applying sample to the next strip.

C. **Running the Electrophoresis.** The instructor will check the wiring and turn on the power supply. The voltage is increased until the milliammeter reading is 2 ma per strip (*e.g.,* if 9 strips are in the chamber, the milliammeter should read 18 ma). Check the milliammeter reading during the first 5-10 minutes until the current stabilizes. The current should never exceed 3.0 ma per strip, otherwise the heat buildup will be excessive and the protein may denature. The voltage should be 300-350 volts, and the current 1-3 milliamperes per strip.

Run the electrophoresis for 45 minutes. Turn off the power switch and *disconnect the electric cord from the A.C. outlet before removing the chamber cover.*

D. **Staining the Strip.** Be sure the power is *OFF.* Remove the Magna-Grips, but do not permit the protein bands on the strip to touch the buffer solution. Transfer the strip horizontally to the staining tray and immerse the strip in the Ponceau S dye solution by pushing it down with a plastic forceps. Place all strips from the chamber in the same staining tray and stain for 5-10 minutes.

Using the plastic forceps, transfer the strips from the staining tray to the washing tray containing 5% acetic acid solution. Succesively rinse the strips 3 to 4 times for one minute with gentle agitation in fresh portions of the acetic acid solution to remove excess and background stain. It is important that all background stain be removed before quantitation procedures are employed. If necessary, these strips may be stored in 5% acetic acid and refrigerated.

At this point the strips may be cleared (part E) for mounting on your report sheet, or they may be eluted (part F) for the determination of relative percentages of the protein fractions.

The identification of the four globulins (alpha 1, alpha 2, beta, and gamma) and the albumin is shown in the illustration on page 202.

E. **Clearing the Strips.** The cellulose polyacetate strip is cleared with 13% glacial acetic acid in methanol to give a cellophanelike appearance. To permit the strip to clear properly, all water must be removed.

Remove the strip from the 5% acetic acid with a forceps, allowing the solution to drain off. Immerse the strip in a tray containing methanol for about one minute. Transfer to a second tray of methanol for another minute to complete the dehydration. Two strips can be processed simultaneously.

Place two dehydrated electrophoresis strips on a clean 10 x 20-cm (4″ x 8″) glass plate. Carefully and slowly immerse the glass plate in a tray containing 13% acetic acid in methanol. (A 6-8 mm glass rod, laying crosswise at one end of the tray, makes for easier removal of the glass plate later.) Adjust the strips with

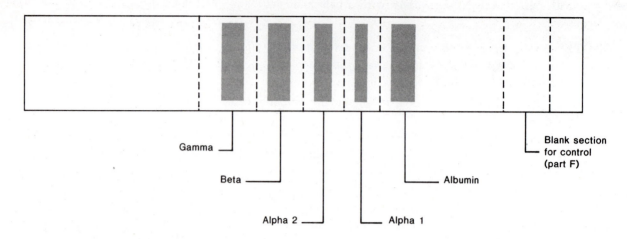

Direction of migration ——————▶

Gamma

Beta

Alpha 2

Alpha 1

Albumin

Blank section for control (part F)

forceps if they slide or raise on the plate. Leave the strips immersed in the clearing solution for one minute and then carefully grasp the elevated end of the glass plate and raise the plate carefully out of the solution, grasping its edges with your fingers. Remove any air bubbles from the strip with a glass rod wetted with clearing solution. This must be done quickly and gently because the strip becomes tacky as it begins to clear and can be easily damaged. Allow excess liquid to drain from the plate.

Place the glass plate with the strips in an oven at 90-100° for 10 minutes, supporting the plate on a test tube block lying crosswise on the oven shelf. The strips become transparent upon drying. When the plate is removed from the oven, lift the end of the *warm* strip with a single-edge razor blade and pull it from the plate. Attach the strip horizontally with Scotch tape across your laboratory report sheet, with the heavy albumin fraction to the right.

F. *Elution and Quantitation of Protein Fractions.* If a strip is to be eluted for quantitation, *it must not be cleared.* After the background stain has been removed by the washings in 5% acetic acid solution, gently blot the strip between clean absorbent pads.

Cut the various protein bands (see dotted lines in illustration at end of part D) and place them in a series of five clean test tubes marked to identify the fractions. A blank area of the strip (see illustration) should be cut also and placed in a sixth tube.

Pipet 3.0 ml of 0.1 N NaOH into each tube. Upon thorough shaking, the dyed protein will completely elute to give a clear solution whose color intensity is proportional to the amount of dye, and hence the amount of protein, in the particular fraction. The polyacetate strips do not dissolve.

NOTE. The 3.0 ml of NaOH is needed to completely intercept the light path in a Spectronic 20. Larger volumes of NaOH may be required for use in other spectrophotometers or colorimeters, but color intensities will be decreased and the accuracy of the determination reduced. It is recommended not to use more than 5 ml of NaOH.

IMPORTANT! *Add 1 drop of glacial acetic acid to each test tube and mix.*Ponceau S dye forms a purple color in NaOH, and the acetic acid converts it back to the more intense red color.

Transfer the solution to a cuvette and measure the absorbance of each solution at 525 nm against water. Read the absorbancies of the blank first and then the globulin fraction, starting with alpha-1 globulin, which should have the lowest absorbance (except for blank). Read the albumin last. In doing this, the cuvette need not be rinsed between fractions, only drained. Rinse the cuvette with water after the albumin reading if the readings of the tubes are repeated. Subtract the absorbance of the blank from the absorbance of each protein fraction only if the blank absorbance is the smallest value. Blanks have high values when all of the background dye has not been removed.

Record your readings and calculate the relative percent of protein in each fraction as indicated on the report sheet.

REFERENCE

Gelman serum protein electrophoresis system. 1978. Technical Bulletin 20. Rev. ed. Ann Arbor, Mich.: Gelman Instrument Company.

Reagents, Materials, and Equipment are listed on page 290.

Name _____ Section _____ Date _____

Experiment 29. ELECTROPHORESIS OF SERUM PROTEINS

Electrophoretic Strip of Protein Fractions After Clearing

Mount the clear electrophoretic strip across the page with the *albumin fraction to the right,* using transparent tape.

Identify each fraction as shown in the illustration in part D.

Above the strip indicate with a long arrow the direction of movement of the protein fractions during electrophoresis.

At the ends of the strip indicate the direct current polarity: $(-)$ cathode and $(+)$ anode.

Voltage and current used: _____ volts _____ milliamperes

Elution and Quantitation of Protein Fractions

From the absorbance of each protein fraction and the blank, calculate the relative percentage of each fraction in the serum.

$$\text{Relative \% of protein fraction} = \frac{\text{Absorbance of fraction}}{\text{Sum of absorbancies}} \times 100$$

Protein fraction	Absorbance of protein fraction	Absorbance of blank	Corrected absorbance	Relative % of protein fractions
albumin				
alpha-1 globulin				
alpha-2 globulin				
beta globulin				
gamma globulin				
Sum of absorbancies				

QUESTIONS

Name _____ Section _____ Date _____

Experiment 29. ELECTROPHORESIS OF SERUM PROTEINS

1. What is the chemical composition of the electrophoresis strip?

   _____

2. What are the constituents and the pH of the buffer? (See Reagents in the Appendix.)

   _____

3. At the pH used in this experiment during electrophoresis, is the net charge on the serum proteins, $+$, $-$, or 0? Explain on the basis of the polar groups of proteins.

   _____

   _____

4. Why is the sample applied closer to the cathode end of the strip?

   _____

5. Give two reasons why the polyacetate strip must be kept moist during sample application and during the electrophoresis process.

   _____

   _____

6. What reagents are used for

   a. Staining? _____   c. Dehydration? _____

   b. Rinsing? _____   d. Clearing? _____

7. Which of the serum protein fractions migrate

   the least? _____   the most? _____

8. Some of the proteins of serum have the following isoelectric points: albumin $= 4.9$, β_1-lipoprotein $= 5.5$, and γ-globulin $= 6.8$. Account for the *relative* ionic charges on these protein fractions and their position on the strip at the pH of the buffer used in this experiment.

Vitamins

Colorimetric Determination of Ascorbic Acid (Vitamin C)

INTRODUCTION

Ascorbic acid is one of the more abundant vitamins and is required in relatively large amounts by humans. It is found in a wide variety of fresh fruits and vegetables but is low or lacking in cereals (unless fortified), fats and oils, and meats except liver. It is a strong reducing agent because of its enediol structure, which is easily oxidized to the corresponding diketo (dehydro or oxidized) form as shown below.

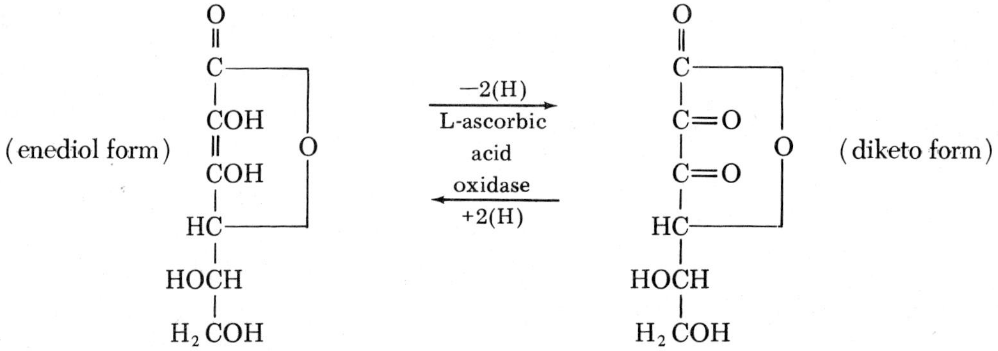

L-Ascorbic acid
(reduced form)

L-Dehydroascorbic acid
(oxidized form)

This reaction, which is catalyzed by copper as well as by the enzyme ascorbic acid oxidase, forms the basis for a method of analysis for vitamin C. The oxidizing agent is a dye that has a sufficiently high oxidation-reduction potential to oxidize the vitamin but does not react with weaker reducing agents. Thus the method really is a measure of certain reducing substances present in fruits and vegetables, of which the main one usually is ascorbic acid. Dehydroascorbic acid, which also occurs naturally in some food materials and has full vitamin C activity, is not measured unless reduced prior to titration with the dye.

In general, fresh foods retain their vitamin C content best when they are protected as much as possible from air. This means that vegetables should be kept within their natural protective coatings as long as possible until just before they are to be eaten; e.g., peas should not be shelled and corn should not be husked until the last moment. Also some vegetables may be kept in containers that keep out the air to a certain degree. Even tightly wrapping in paper or plastic will help considerably. The food sample should be kept as cold as possible without freezing. Refrigerator storage is preferable to room temperature storage.

PROCEDURE

A. *Titration of Dye with Ascorbic Acid.* The standardization of the dye, 2,6-dichloroindophenol sodium salt, will give you the opportunity of observing the titration endpoint before you analyze a fruit or vegetable sample.

Obtain a 10-ml buret from the storeroom or instructor and clean it thoroughly with water, allowing it to drain completely. Note that the smallest division on the graduated scale is 0.05 ml. In a clean, *dry 50-* to 100-ml beaker, obtain about 25 ml of standard ascorbic acid (0.1 mg/ml) in 1% oxalic acid. Rinse the buret with several 2-3-ml portions of the ascorbic acid solution, and then fill the buret properly.

In order to avoid waste, the expensive dye solution is dispensed from a 10-ml buret on the reagent bench. The buret can easily be refilled from a polyethylene "squeeze bottle." It is very important to measure the *exact* volume of the dye to obtain accurate and reproducible analytical results.

Drain exactly 2.00 ml of the dye from the buret into a scrupulously clean 100-ml beaker and directly titrate it with the standard ascorbic acid until the red color of the dye just disappears. (Swirl the beaker by hand.) This usually requires about 4.5-6.5 ml of standard ascorbic acid. Use a white background and a water blank to better observe the endpoint. Repeat titrations until your results check within 0.1 ml.

Calculate the weight of ascorbic acid oxidized by 2.00 ml of dye solution as follows:

$$2 \text{ ml dye} = \text{ml ascorbic acid} \times 0.1 \text{ mg/ml}$$

The reaction is an oxidation of ascorbic acid and the reduction of the dye. See the introduction for reduced and oxidized structures of ascorbic acid.

NOTE: The solution *must* be acidic with oxalic acid for proper color changes to occur. That is why the standard ascorbic acid and samples are prepared with 1% oxalic acid.

B. *Titration of Dye with Prepared Samples of Fruits and Vegetables.* The fruits and vegetable samples prepared in 1% oxalic acid as described in part C will be used in place of the standard ascorbic acid to titrate 2.00 ml of the dye solution until the red color disappears as you observed in part A. The endpoint will be more difficult to see, because the prepared sample usually has a slight interfering color, and larger volumes of the sample solution will have been added to reach the endpoint.

The prepared sample solution is transferred to a 50-ml buret and is then used to titrate exactly 2.00 ml dye in a 100- or 150-ml beaker. The volume of the sample solution required to completely decolorize the dye contains the *same amount of ascorbic acid* as is present in the volume of standard ascorbic acid used to reach the endpoint in part A. From this relationship the vitamin C content of the original fruit or vegetable can be calculated.

C. *Preparation of Samples.* Check the bulletin board for your assignment of fruits or vegetables. If sufficient volume of the prepared extract is available, run duplicate titrations.

(1) *Citrus fruit juices.* The juice from fresh fruits can be expressed with a lemon squeezer, but avoid getting too much pulp or fibrous material. Strain through several folds of cheesecloth. Canned fruit juices or reconstituted concentrates should be filtered through cheesecloth or glass wool to remove fibrous materials that might plug the tips of pipets and burets.

Pipet 5 ml (5 g) of the fruit juice into a 50-ml volumetric flask. Dilute to the mark with *1% oxalic acid* and mix thoroughly. Rinse a 50-ml buret with 2-3 ml of this solution and discard. Pour the remainder of the solution into the buret and titrate 2.00 ml of dye solution as described above.

(2) *Vegetables.* In a Waring blender homogenize 5 g (weighed to 0.01 g) of cabbage, lettuce, fresh spinach, green pepper, string beans, raw potato, or other vegetable with 10-15 ml of 1% oxalic acid. Pour off the liquid onto a filter paper and collect the filtrate in a 100-ml graduate. Repeat with a second 10-15-ml portion of the acid and filter. Add 1% oxalic acid to the filtrate until the volume is exactly 50 ml, mix, transfer to a buret, and titrate against the dye as described above.

(3) *Destruction of vitamin C by boiling.* To 5 g of cabbage or other vegetable in a beaker, add water and boil for as long as would be required to cook it for table use. Drain off the cooking water, dilute to 25 or 50 ml *with 1% oxalic acid,* and titrate against the dye. The 50-ml volume may be too dilute for satisfactory titration if little vitamin C is present. If too dilute, use 0.50 or 1.00 ml of dye.

Transfer the boiled vegetable to the Waring blender, homogenize twice with 1% oxalic acid and filter as in (2) above. Dilute to 25 or 50 ml with 1% oxalic acid, and titrate against the dye using 0.50 or 1.00 ml of dye.

D. *Independent Vitamin C Stability Experiment.* Each student individually is to obtain a sample of some kind of food, such as a fresh fruit or vegetable, and study the loss of vitamin C from this food that results from ordinary cooking or other common procedures to which such foods are frequently subjected in the home.

For this purpose the student should:

1. Obtain the food sample. If this is not convenient, ask the laboratory instructor to order whatever is needed. In general, 183 grams of the food sample should be sufficient.
2. Divide it into two approximately equal portions.
3. Keep one portion under conditions best calculated to preserve the vitamin C content as explained in the last paragraph of the introduction to this experiment.
4. Subject the other portion to the desired treatment.
5. Bring both portions to the laboratory and determine the vitamin C content according to the procedure given in part C. Consult with the laboratory instructor as to the exact details of how you should prepare the extract of your samples for the vitamin C analyses. In general this will be done as described in part C (1) and (2).

Types of treatments to be studied:
1. Making a "rosebud" out of a radish.
2. Boiling vegetables, e.g., string beans, in water until tender and draining off the water.
3. Peeling and slicing of fruit, such as an apple or banana, and letting it stand 10 or 15 minutes at room temperature.
4. Effect of mashing potatoes as compared to baking or boiling.
5. Effect of storing several different types of foods at room temperature as compared to storing in the refrigerator for several days, e.g., oranges, tomatoes, lettuce, or string beans.
6. Comparing fresh, frozen, and canned samples of the same vegetable or fruit other than those already used in part C.
7. Comparing liver with muscle meat as a source of vitamin C.
8. Comparing the most expensive type of animal liver, such as calves' liver, with the least expensive types, such as pork liver, as sources of vitamin C.
9. Any other similar treatment or comparison which the student wishes to study, provided approval of the instructor is secured first.

REFERENCES

HARRIS, L. J., and OLLIVER, M. 1942. *Biochem. J.* 36:155.
ZEPPLIN, M., and ELVEHJEM, C. A. 1944. *Food Research* 9:100.
WATT, B. K., and MERRILL, A. L. 1963. *Composition of foods.* U.S. Dept. of Agr. Handbook No. 8, Washington, D.C.: Supt. of Documents, U.S. Government Printing Office.

Reagents and Materials are listed on page 291.

Name _____ Section _____ Date _____

Experiment 30.   COLORIMETRIC DETERMINATION OF ASCORBIC ACID (VITAMIN C)

Standardization of Dye

Conc. of standard ascorbic acid = 0.1 mg/ml

_____ ml ascorbic acid (vitamin C) to decolorize 2 ml dye solution.

_____ mg vitamin C equivalent to 2 ml dye.

Calculations and Results

$$\frac{\text{mg vitamin C equiv. to 2 ml dye} \times 100}{\text{ml sample ext. for titration} \times \text{g sample equiv. to 1 ml ext.}} = \text{mg vit. C/100 g sample}$$

Sample Data						
Name of juice or vegetable	weight	volume to which diluted	sample equiv. to 1 ml ext	vol. to titrate dye	2 ml dye vitamin C equivalent	vitamin C found in sample
	grams	ml	grams	ml	mg	mg/100 g

Destruction of Vitamin C by Boiling

Name of boiled vegetable_____

Vitamin C found: (Record titration data in table above.) If less than 2 ml dye is used in the titration, substitute the correct vitamin C equivalent in the table and the equation above.

in raw sample _____ mg/100 g sample

in boiled sample_____mg/100 g sample

in cooking water_____mg/100 g sample

Loss of vitamin C by cooking:

actual destruction ...._____%   overall loss assuming cooking water discarded ...._____%

D. *Independent Vitamin C Stability Study.* Write up the experiment so as to show (a) just what the sample food was, when and where it was obtained, and in what way and for how long it was subjected to the treatment being studied; (b) the results of the vitamin C determinations on each portion of the food, including all experimental values, such as the amount of sample used, the volume to which it was diluted, the number of ml needed for titration of the dye, number of ml of dye used, and final calculation of the vitamin C content per 100 g of food for each sample; and (c) conclusions as to how much vitamin C was lost, what percentage this was of the original vitamin content, how this food should or should not be handled in relation to its vitamin C, and also a brief discussion of the value of this particular food as a practical source of vitamin C in our diets in this locality.

Name_____Section_____ Date _____

Experiment 30. COLORIMETRIC DETERMINATION OF ASCORBIC ACID (VITAMIN C)

1. Show with *partial structures* the part of ascorbic acid that is oxidized and that part of the dye that is reduced when the two compounds react.

2. What are the chief sources of error in this method of determining vitamin C?

_____

_____

3. If the entire volume of an extract (e.g., the cooking water) fails to decolorize the dye, what should you do to complete the titration so you still can obtain your desired result?

_____

_____

4. Using data obtained from your own results, from the results of other members of your class, or from reference books, calculate the percentage loss of vitamin C, if any, that occurs when fresh orange juice is either canned or converted to frozen concentrates and later reconstituted. Assume 100 ml = 100 g.

orange juice	mg vit. C / 100 ml	% loss, if any
raw	_____	0.00
canned	_____	_____
frozen (reconstituted)	_____	_____

5. From the vitamin C content of the fresh fruits and raw vegetables listed below, calculate the relative cost of vitamin C in these foods. Use current market prices. (Vitamin C values taken from "Composition of Foods," USDA Handbook No. 8, Table 2)

	mg vit. C per lb. of food as purchased	cost of food per pound	relative cost expressed as cents/mg vitamin C
grapefruit	85	_____	_____
orange	166	_____	_____
cabbage	192	_____	_____
green pepper	476	_____	_____
potato	73	_____	_____
tomato	102	_____	_____

Extraction and Chromatography of Carotenes. Provitamin A

INTRODUCTION

The carotenes belong to a group of terpenes called carotenoids. The unsaturated hydrocarbon molecules of these terpenes consist of eight isoprene units. Four of the isoprene units make up two methyl-substituted cyclohexene rings, which are linked together by a linear chain of the remaining four isoprene units. They occur widely as plant pigments, and they are mainly responsible for the orange color of carrots. The most important and abundant carotene is beta-carotene, which is enzymatically converted in the body into two molecules of vitamin A. The symmetrical β-carotene is cleaved into two identical parts, and the terminal carbon of the unsaturated side chain is oxidized to a primary alcohol ($-CH_2OH$).

$$\beta\text{-carotene} \xrightarrow[\text{intestines}]{\text{in liver and}} 2 \quad \overset{CH_3 \quad CH_3}{\underset{CH_3}{\bigcirc}} \overset{CH_3}{\underset{|}{\text{CH=CH-C=CH-CH=CH-}}} \overset{CH_3}{\underset{|}{\text{C=CH-CH}_2\text{OH}}}$$

Provitamin A   Vitamin A₁ or Retinol

In this experiment a mixture of carotenes extracted from carrots will be separated on a column by adsorption chromatography.

Adsorption chromatography is a procedure by which a mixture of closely similar chemical substances may be separated into its individual components and each obtained in purified form. In this experiment a mixture of carotenes extracted from carrots will be separated by this method.

A solution of the sample mixture is poured into a tube containing a narrow column of finely-powdered adsorbent, which retains the mixture on or near the surface of the column. The chromatogram is then "developed" by slowly running through the column a solvent that can elute (desorb) the sample, at least partially, and therefore move it a short distance down the column. Repeated adsorptions and elutions occur under these circumstances. If the individual components of the mixture differ even slightly in the ease with which they are adsorbed or eluted, some of them will move down the column faster than others. This results in the formation of distinct bands, each of which under favorable circumstances contains one of the components of the original mixture.

The carotenes are named on the basis of the rate at which they flow through such a column. Alpha carotene flows through most rapidly followed in order by beta and gamma. The carotene from carrots is predominantly beta, hence this band is the broadest of the three. Any traces of carotenols will appear above the band of γ-carotene.

Chromatography finds numerous applications in the purification and isolation of many types of compounds, both colored and uncolored, including plant pigments, sterols, tocopherols, hormones, vitamins, enzymes, etc. In each case success is dependent upon the proper selection of adsorbent and solvent. The desired compound may be isolated by allowing it to flow from the bottom of the column or by extruding the column and mechanically separating the desired band. In the latter case, a more polar solvent than the developing solvent will remove the compound from the adsorbent.

It is usually necessary to apply a gentle vacuum at the bottom of the column in order to obtain a satisfactory rate of flow of the developing solvent. If, however, the adsorbent is extremely finely divided, it may be necessary to mix it with some coarser, inert, and nonadsorbing material in order to achieve a satisfactory flow rate. A purified, diatomaceous earth preparation marketed under the trade name "Celite" (Johns-Manville Company) is useful for this purpose.

PROCEDURE

The experiment can be performed as a class demonstration, or the class may be divided into groups of 2-4 students and each group can do it under close supervision of the instructor. The heating of hot alcohol during the extraction of carrots and the procedures involving the use of petroleum ether should be carried out by the appropriate safe methods. No lighted burners in the area!

A portion of the prepared extract can be used for the separation of the carotenes by thin-layer chromatography (part C).

A. **Preparation of Crude Carotene Extract.** Extract 20 g of finely grated carrots with 200 ml of hot 95% ethanol in a 500-ml Erlenmeyer flask for half an hour. Decant the yellow solution, dilute to approximately 85% ethanol with water, and cool to room temperature. Shake in a separatory funnel with 50 ml of petroleum ether (Skellysolve B, b.p. 60-70°C). On standing, two layers will separate, an upper one of petroleum ether carrying the carotenes, and a lower one of ethanol carrying the xanthophylls. Separate the two layers and extract the alcohol solution repeatedly with small amounts of petroleum ether until the upper layer remains colorless. Wash the petroleum ether extracts once with 85% ethanol, to remove possible traces of xanthophylls. Concentrate the petroleum ether layer *in vacuo* just to dryness to remove traces of polar solvents. Dissolve the oily residue in 5 ml petroleum ether and preserve in as small a container as possible for chromatographing. To minimize oxidation of the carotene keep it in a tightly stoppered container in the dark. (NOTE: This preparation should be carried out by the instructor.)

B. **Column Chromatography.** Carefully pack a column of 1:1 Celite-MgO adsorbent to a depth of about 7.5 to 10 cm (3-4″) in a 11-mm (i.d.) chromatographic column with a stopcock. To do this, push a small plug of glass wool into the bottom of the column. Pour about 2.5 cm of the adsorbent mixture into the column and tamp it gently with a flat-end 10-mm glass rod. Add repeated small portions of adsorbent and tamp lightly until the 7.5- to 10-cm height is attained. The instructor will demonstrate the technique. When the column is packed, add a few ml of petroleum ether to the column in such a way that the surface of the column is not disturbed. Once solvent is added, it is essential that the top of the column remain covered with liquid. Gently pour a 1-cm layer of anhydrous Na_2SO_4 on top of column. This protects the column from traces of water and prevents mechanical erosion. Just before the last of the solvent has been drawn into the column, pipet 1 to 3 ml of the carotene solution onto the column. Draw most of it in, then wash the balance quantitatively onto the column with 2- or 3-ml portions of petroleum ether. *When there is little color in the liquid,* add a larger amount of the developer and allow the development to proceed. If the column has been properly packed and all traces of polar solvents are absent, well-developed bands of color, separated by clear white bands of the adsorbent, should appear. A flow rate of 1 ml/minute gives good separation.

C. **Thin-Layer Chromatography.** The petroleum ether extract can be used for the separation of the carotenes by thin-layer chromatography. Refer to Experiment 21 for details. A 0.05% solution of β-carotene in petroleum ether can be applied as a reference. Suggested solvent systems are: (1) petroleum ether, benzene (50:50 v/v), and (2) the system used in Experiment 21. The color spots can be intensified and protected by spraying the chromatogram with 5% paraffin oil in petroleum ether. Attach your chromatogram with transparent tape on your report sheet if the precoated chromatography sheets are used.

REFERENCE

STAHL, E., ed. 1969. *Thin-layer chromatography.* 2d ed. New York: Springer-Verlag New York Inc. pp. 376-79.

Reagents and Materials are listed on page 291.

LABORATORY REPORT

Name _____ Section _____ Date _____

Experiment 31.   EXTRACTION AND CHROMATOGRAPHY OF CAROTENES. PROVITAMIN A

B. *Column Chromatography*

Draw a sketch of the chromatographic column. Show the location of the bands and identify each. Note particularly the relative width and color intensity of each band.

C. *Thin-Layer Chromatography*

Solvent System _____

	Millimeters	R_f *value*	*Attach Chromatogram*
Solvent front	_____	__1.00__	
α-carotene	_____	_____	
β-carotene	_____	_____	
γ-carotene	_____	_____	
Reference β-carotene	_____	_____	
Other carotenoids	_____	_____	

Name_____ Section _____ Date _____

Experiment 31. EXTRACTION AND CHROMATOGRAPHY OF CAROTENES. PROVITAMIN A

1. Based on its empirical formula, $C_{40}H_{56}$, to what
   class of organic compounds does β-carotene belong?_____

2. The structure of β-carotene can be divided into eight repeating 5-carbon units. What is the name and structure of this repeating unit and to what class of compounds do the carotenes belong based on this unit?

   Name of unit_____   Structure of unit.

   Name of class of compounds. _____

3. Write the structure of β-carotene and indicate with an arrow the point of cleavage when it is physiologically cleaved to vitamin A_1.

4. In preparing the carotene extract from carrots
   a. Why are the carrots first extracted with 95% ethyl alcohol instead of immediate extraction with petroleum ether?

   _____

   _____

   b. Why would you expect carotenes to be more soluble in petroleum ether than in 85% ethanol?

   _____

5. Name several other adsorbents commonly used in column chromatography.

   _____

6. What is the purpose of the Celite in the Celite-MgO mixture?

   _____

Enzymes and Metabolism

GENERAL INTRODUCTION

The chemical reactions that go on in living organisms and are essential for growth, reproduction, movement, and all other vital functions are individually and collectively referred to as *metabolism*. Many thousands of such metabolic processes are occurring constantly in even the simplest forms of life.

Enzymes are proteins that act as catalysts for metabolic reactions. They increase the rate of the reaction, but do not influence the kind or amount of products formed. In general, each metabolic reaction has to be catalyzed in the living organism by its own special enzyme.

The existence of enzymes in biological materials can be demonstrated by the effects they bring about. In this section digestive enzymes capable of breaking down food products, such as carbohydrates, fats, and proteins, will be studied, as well as enzymes involved in energy metabolism in yeast, and in plant, insect, and animal tissues. Enzymes can also be recognized by the fact that their characteristic activities are destroyed by heating to 100°, by extreme pH changes, or by various chemical inhibitors and poisons.

Food materials must be broken down by digestion before they can be absorbed and utilized by the body. Consequently if an organism has no enzyme capable of attacking a particular type of carbohydrate, protein, or fat, that particular substance will have no food value for the organism. Carbohydrates and fats, as we have seen earlier, can also be broken down by strong acids or alkalies. However, digestive enzymes can accomplish the same breakdown under conditions compatible with life, that is, at moderate temperatures and physiological pH ranges.

Enzymatic Digestion of Proteins. Factors Affecting Enzyme Activity

INTRODUCTION

The hydrolysis of peptide bonds that join the amino acid residues of proteins and polypeptides is catalyzed by several proteolytic enzymes, such as trypsin, chymotrypsin, and carboxypeptidase. These particular proteinases are derived from their respective inactive proenzymes, or zymogens, that are present in the pancreatic juice. Upon entering the duodenum of the small intestine, the zymogens are converted to their active forms by the splitting off of a small polypeptide fragment. Once formed, trypsin itself catalyzes the conversion of trypsinogen, chymotrypsinogen, and procarboxypeptidase to active hydrolases. Proteins are digested by these enzymes to small polypeptides and amino acids. In their catalytic role, these proteinases display some selectivity for certain peptide bonds, depending on which amino acids are involved in that bond. Thus the digestive actions of these enzymes are complementary and, together with the aid of other intestinal enzymes, amino acids are the final product. The commercial animal pancreatin powder used in this experiment is a good source of trypsin.

The substrate for this experiment is the gelatin layer on an exposed and developed photographic film. When the gelatin is hydrolyzed by the proteolytic enzyme, the black, reduced silver particles will flake away from the plastic backing of the film in approximate proportion to the degree of digestion of the gelatin. If sufficient digestion has occurred, the film backing becomes transparent; if no digestion occurs, the film negative retains its black color.

The rate of activity of enzymes is dependent on such factors as concentration of enzyme, temperature, pH, and inhibition. The effect of these factors will be noted in the experiments below.

PROCEDURE

NOTE: The pancreatin extract solution and the various buffer solutions will be available in 50-ml burets on the reagent bench.

A. *Effect of Enzyme Concentration.* Into each of four test tubes add the solutions indicated.

Tube No.	Pancreatin extract ml	pH 8.0 buffer ml	Water ml
1	3.0	3.0	0
2	1.5	3.0	1.5
3	0.5	3.0	2.5
4	0	3.0	3.0

Allow all tubes to attemper to 40°C in a water bath for several minutes. Note the time (zero time) and then quickly drop a 8 x 12-mm film strip into each tube. Carefully rotate each tube to ensure complete immersion of the film. The tubes may be very gently rotated occasionally, but *do not shake* them. When the strip in tube 1 is almost clear, immediately add an equal volume of dilute (1:1) formaldehyde solution (18-19% HCHO) to each tube. Slowly pour off the solution from each tube, being careful to retain the film, and just cover the film with fresh, dilute formaldehyde solution and allow the tubes to stand for 10 minutes. Again, carefully pour off the formaldehyde solution, and wash each strip several times with fresh portions of water.

CAUTION: Formaldehyde will cause irritation and shriveling of skin. Avoid spillage on fingers. Eye irritation may occur in areas of poor ventilation. Use under a fume hood when possible.

Place the strips with emulsion (dull) side up on a paper towel to dry. Permanently mount the dried film strips with transparent tape on your laboratory report sheet. Properly identify each by noting the ml of pancreatin extract used for the digestion of gelatin.

B. *Effect of pH.* Add 3 ml of pancreatin extract into each of four test tubes. Then add 3 ml of a phosphate buffer having a pH as follows: tube 1, pH 5.0; tube 2, pH 6.0; tube 3, pH 7.0; and tube 4, pH 8.0. Let all tubes attemper to 40°C, note the time, and add a film strip to each and check for complete immersion. When one strip is almost clear (but need not be complete), add formaldehyde to all tubes and continue as in part A.

C. *Effect of Temperature.* Into each of four tubes, add 3 ml of pancreatin extract and 3 ml of pH 8 buffer solution. Place tube 1 in a boiling water bath for two minutes, cool back to about room temperature and place it, together with unboiled tube 2, in a water bath to attemper to 40°C. Also place tubes 3 and 4 in an ice-water bath and allow them to reach 0°C with occasional shaking. Add a film strip to each tube and check for complete immersion. When one strip shows definite evidence of clearing, add an equal volume of formaldehyde to tubes 1, 2, and 3 *only*. Place tube 4 into the 40° bath until it begins to clear, then treat it similarly. All strips are further treated as in part A.

D. *Effect of Inhibitor.* Into each of two test tubes add 3 ml of pancreatin extract and 3 ml of pH 8 buffer solution. To one of the tubes add 2 drops of 0.2 M mercuric chloride. Attemper the tubes to 40°C and proceed as in part A.

REFERENCES

Gates, F. L. 1926-27. *Proc. Soc. Exp. Biol. Med.* 24:936.
Harrow, B.; Borek, E.; Mazur, A.; Stone, G. C. H.; and Wagreich, H. 1965. *Laboratory manual of biochemistry.* 5th ed. Philadelphia: W. B. Saunders Co., p. 70.

Reagents, Materials and Equipment are listed on page 292.

Name_____ Section _____ Date_____

Experiment 32.   ENZYMATIC DIGESTION OF PROTEINS. FACTORS AFFECTING ENZYME ACTIVITY

Mount the film strips under the appropriate headings with a single strip of transparent tape for each set. Identify each strip with conditions used during digestion.

A. *Effect of enzyme concentration*

B. *Effect of pH*

C. *Effect of temperature*

D. *Effect of inhibitor*

Based on the above evidence, summarize your conclusions concerning the effect of these four factors on the digestion of gelatin.

A. Effect of enzyme concentration.

B. Effect of pH

C. Effect of temperature.

D. Effect of inhibitor.

Name_____ Section _____ Date_____

Experiment 32.   ENZYMATIC DIGESTION OF PROTEINS. FACTORS AFFECTING ENZYME ACTIVITY

1. What is the actual substrate in this experiment?_____

2. Why is the photographic film exposed and developed before use?

    _____

3. Which enzyme in the *intestinal* juice catalyzes
    the conversion of trypsinogen to trypsin?   _____

4. What other enzymes in *pancreatic* juice are activated by trypsin?

    _____

5. Using E and S to represent enzyme and substrate, respectively, write a general equation for the mechanism of an enzyme-catalyzed reaction.

    _____

6. What is the purpose of adding formaldehyde at the end of the digestion period in this experiment?

    _____

7. a. In the space to the right, draw
        and label a graph showing the
        relationship between increasing
        enzyme concentration and the
        velocity of reaction (products
        produced or substrate con-
        sumed).

    b. How did you qualitatively show this relationship in this experiment?

        _____

8. a. Sketch and label a graph show-
        ing the relationship between ve-
        locity of reaction and pH.

    b. What is the optimum range for trypsin and most other pancreatic enzymes?   _____

    c. What is the optimum pH range for pepsin?   _____

    d. What is the optimum pH range for *most* enzymes in the body?   _____

9. a. Sketch and label a graph showing the relationship between ve-
locity of reaction versus temperature.

b. Explain the effect of low temperature and of high temperature on an enzyme reaction as you observed it in this experiment.

_____

_____

10. a. What active site on enzymes can be inhibited by Hg^{++}?_____

b. What other heavy metal ion inhibits this site? _____

c. List several other ions or substances that poison enzymes. _____

_____

11. What factors, in addition to those studied in this experiment, can influence enzyme activity?

_____

_____

12. Name two enzymes, and their sources, that catalyze the hydrolysis of short-chain polypeptides to free
amino acids.

_____

_____

13. In what way could the biuret test (Expt. 24) be useful in following the digestion of proteins?

_____

_____

Hydrolysis of Starch by Salivary Amylase

INTRODUCTION

Amylases are enzymes that catalyze the hydrolysis of the $\alpha,1\rightarrow4$ glucosidic linkages of certain polysaccharides, such as glycogen and the amylose and amylopectin components of starch. The salivary alpha amylase, ptyalin, begins the digestion of these carbohydrates in the mouth. The digestion continues briefly in the stomach until the pH drops too low, and is completed in the intestines by the attack of another alpha amylase. Amylases do not catalyze the hydrolysis of the $\alpha,1\rightarrow4$ linkages joining the galacturonic acid units in pectin because the amylases are glucosidases. In fact, pectin is a polysaccharide that is nondigestible and is commonly used in the control of diarrheal diseases. (See reference.)

The term *alpha and beta* amylases (plant enzymes) can be very confusing, since they both catalyze the hydrolysis of $\alpha,1\rightarrow4$ glucosidic linkages, but not $\beta,1\rightarrow4$. The β-amylases occur naturally in the cereal grains (e.g., ungerminated barley), while the α-amylases are produced during the germination of grain seeds in the "malting process."

The β-amylases are capable of splitting off maltose from the nonreducing ends of amylose, amylopectins, and dextrins. Hence, the β-amylases are often referred to as "saccharifying" enzymes. The α-amylases attack amylose and amylopectin of starch *at random,* producing maltose and various lower molecular weight dextrins. Alpha-amylases hydrolyze $\alpha,1\rightarrow4$ linkages between the $\alpha,1\rightarrow6$ branch points in amylopectin, something that β-amylases are incapable of doing. By their action, α-amylases readily reduce the viscosity of starch suspensions, thus they are also called "liquefying" amylases. Malted barley, which is commercially produced in large quantities, contains a mixture of alpha and beta amylases.

PROCEDURE

Hydrolysis of Starch. Collect 5-10 ml of saliva in a beaker. Place 10 ml of 1% starch solution in a test tube. With a stirring rod, transfer 1 or 2 drops of the starch to a depression in a spot plate and add a drop of iodine solution. Note the color produced before addition of saliva to the starch solution.

Add 3 drops of saliva to the starch solution and mix with the stirring rod. Set the tube in a 40° water bath and note the time. Do *not* overheat or you will inactivate the enzyme. Remove a drop of the mixture in 2 minutes and immediately test with iodine solution and note the color. Repeat the iodine test at 5 minutes, and at 5-minute intervals thereafter. When no color other than that of the iodine solution is produced, apply Benedict's and Barfoed's tests to 1-2-ml portions of the hydrolyzed starch solution as instructed in Experiment 10.

NOTE: If no reaction-color is produced with iodine after the first 5 minutes of the hydrolysis of starch, repeat the experiment with a fresh 1% starch solution, but with 1-2 drops of saliva so that you can observe the iodine-color changes. On the other hand, if the blue color of the starch-iodine reaction persists after 15 minutes, start over with fresh starch solution and about 8 drops of saliva.

REFERENCES

CHENOWETH, W. L., and LEVEILLE, G. A. 1975. *Physiological effects of food carbohydrates*, ed. A. Jeanes and J. Hodge. ACS Symposium Series 15. Washington, D.C.: American Chemical Society, p. 312.

JEANES, A. *Ibid.* p. 336.

Reagents and Materials are listed on page 294.

Name_____ Section _____ Date_____

Experiment 33.   HYDROLYSIS OF STARCH BY SALIVARY AMYLASE

List the color changes you observed when applying the iodine test at the time intervals indicated until no color other than the iodine color was produced.

Minutes	*Color-Iodine test*	*Minutes*	*Color-Iodine test*
(Before saliva)	_____	10	_____
2	_____	15	_____
5	_____	20	_____

What results were obtained when applying the following tests to portions of the hydrolyzed starch solution?

Benedict's Test_____

Barfoed's Test_____

What products are formed by complete enzymatic hydrolysis of starch by α-amylase?

_____   _____

QUESTIONS

Name_____ Section _____ Date_____

Experiment 33. HYDROLYSIS OF STARCH BY SALIVARY AMYLASE

1. Write word equations showing the stepwise hydrolysis of starch by salivary amylase. Name the products formed at each step and the relative color of the iodine test.

2. How do the final products of the above hydrolysis differ from that produced by heating starch with dilute HCl (Expt. 10)?

   _____

   _____

3. Give the common names for the following amylases:

   Salivary amylase_____   Pancreatic amylase_____

4. How do the "alpha" and "beta" amylases derived from cereal grains compare as to the following:

   a. Type of linkages hydrolyzed _____

   b. Source of each_____

   c. Mode of action on amylopectin and glycogen.

      α-amylase _____

      _____

      β-amylase _____

      _____

5. Fill in the following table:

Polysaccharide	Glycosidic linkage present	Susceptibility to digestion by amylase	Food value for man
Dextrin			
Amylose			
Cellulose			
Dextran			
Amylopectin			
Glycogen			
Pectin			

230

Enzymatic Digestion of Fat

INTRODUCTION

The hydrolysis of triacylglycerols (triglycerides) by pancreatic lipase yields, primarily, free fatty acids and 2-monoacylglycerols with some diacylglycerols and very little glycerol. (NOTE: A fatty acid esterified to the hydroxyl on carbon-2 of glycerol is called a 2-monoacylglycerol.) The accumulation of the free fatty acids gradually reduces the pH of the fat mixture, as can be noted by the presence of an indicator. Most of the fatty acids present are of the C-16 and C-18 variety. Milk fat, however, contains several percent of the C-4 and C-6 fatty acids, and the student should note any change in odor after the hydrolysis of the fat.

Bile salts are important in the digestion of fats because they lower the surface tension of the fatty material, which enables them to be emulsified in an aqueous medium. As emulsifiers, they greatly increase the surface area of the fatty substrate and hasten the hydrolysis by increased enzyme catalysis.

PROCEDURE

A. *Milk Fat Digestion by Pancreatic Lipase.* To about 10 ml of fresh milk add 8-10 drops concentrated litmus solution to impart a deep color, and 1-2 drops 10 percent Na_2CO_3 to produce a distinctly blue shade (pH 7). Divide the blue milk into equal portions and transfer them to test tubes. To one add 3 ml of *boiled* pancreatin solution, and to the other 3 ml of *unboiled* pancreatin solution. Shake each mixture thoroughly. Rinse the *outside* of the test tubes with water to remove any milk so that the laboratory water bath is not fouled. Place the tubes in water at 40-50° C until the end of the laboratory period. Shake the tubes every 20-30 minutes and note any change in color and odor.

B. *Effect of Bile Salts on Lipase Action.* Prepare a set of four digestion and two control mixtures in test tubes as follows, adding the pancreatin solution last:

Tube No.	Vegetable oil ml	Sodium cholate solution ml	Water ml	Pancreatin solution ml	
1	0.5	0	4.5	5	
2	0.5	0	4.5	5	
3	0.5	1	3.5	5	
4	0.5	1	3.5	5	
5	0	0	5	5	Control for 1, 2
6	0	1	4	5	Control for 3, 4

Stopper each tube, shake vigorously for 10-15 seconds, place them all at the same time into a 40° C water bath and incubate at this temperature for 1 hour. Shake the tubes vigorously every 10-15 minutes. Then add 10 drops of 0.5% phenolphthalein indicator solution and titrate with approximately 0.02 N NaOH to a distinctly pink end point color which lasts at least 30 seconds. Titrate tubes 1 and 2 first and use the endpoint color in these tubes as a standard for titration of the others. Record the titration volumes and exact normality of the NaOH in part B of the laboratory report.

Reagents, Materials, and Equipment are listed on page 294.

LABORATORY REPORT

Name_____ Section _____ Date _____

Experiment 34.   ENZYMATIC DIGESTION OF FAT

A. *Milk Fat Digestion by Pancreatic Lipase*

What changes did you observe by the end of the laboratory period? Explain the cause of a color change or the lack of a color change._____

Boiled pancreatin:_____

_____

Unboiled pancreatin:_____

_____

Milk fat and butter should have a characteristic odor after hydrolysis. What causes this odor? Did your test solution have an odor change?

_____

Using oleopalmitobutyrin to represent butterfat, write an equation for the change catalyzed by the enzyme leading to the formation of the 2-monoacylglycerol. Use condensed structural formulas for the fatty acids involved.

B. *Effect of Bile Salts on Lipase Action.* From the titration data in the table below, determine the average volumes of NaOH consumed by tubes 1 and 2 and tubes 3 and 4. Correct the average values by subtracting the ml NaOH required by their respective controls. Calculate the milliequivalents of free fatty acids produced with and without bile salts and compare the relative amounts in the last column.

Normality NaOH_____N

Tube no.	NaOH ml	Average ml	Corrected ml	Free fatty acids	
				meq	ratio
1					
2					
3					
4					
5			Control for 1 and 2		
6			Control for 3 and 4		

QUESTIONS

Name_____ Section_____ Date_____

Experiment 34. ENZYMATIC DIGESTION OF FAT

A. *Milk Fat Hydrolysis by Pancreatic Lipase*

    1. a. Briefly outline the steps in the preparation of "Unboiled pancreatin." (See Reagents in the Appendix)

_____

_____

        b. Give two reasons for adjusting the above pancreatin solution to pH 8.

           1._____

           2._____

    2. a. Name two human enzymes that bring about the digestion of fats and indicate their source.

           1._____

           2._____

        b. Which one is most effective? Why?_____

_____

    3. What color are *pure* fatty acids?

        Solid_____   Liquid _____

    4. Name the fatty acids that are present in butter (milk fat) that are not present in most other natural fats.

_____

    5. What gives butter its natural color?_____

        Why is June butter more highly colored?_____

B. *Effect of Bile Salts on Lipase Action*

    1. Where is bile synthesized and where is it stored?

_____

    2. Name two bile salts other than sodium cholate.

    _____   _____

    3. Show the structure of sodium cholate.

Fermentation of Sugars

INTRODUCTION

The utilization of glucose by yeast cells follows the same metabolic pathway as that by the mitochondria of animal cells with the exception of the conversion of pyruvate to ethanol. The same enzymes, coenzymes, and cofactors catalyze the biochemical reactions involving the hexose phosphates and the triose phosphates. In the final steps, however, pryuvate is decarboxylated to acetaldehyde, which in turn is reduced to ethanol in a reaction catalyzed by alcohol dehydrogenase (ADH) in the presence of the coenzyme, reduced nicotinamide adenine dinucleotide (NADH). The NAD^+ formed is recycled by a coupled reaction for reuse in the oxidation of glyceraldehyde-3-phosphate to 1,3-diphosphoglycerate.

In addition to glucose, other sugars will be tried to determine if the yeast that is used contains the enzymes to convert these sugars to the hexose phosphates required for the glycolytic pathway. Not all types of yeasts have the same activity. For example, brewer's yeast contains maltase for the hydrolysis of maltose to glucose, while certain varieties of baker's yeast lack this activity.

The detection of carbon dioxide or ethanol is evidence that the fermentation of a sugar has occurred. The fermentation tube traps the evolved gas in the closed end of the tube, and a reaction with added NaOH solution confirms the presence of CO_2. The iodoform test on a fermented glucose solution shows the presence of ethanol, which is oxidized to acetaldehyde by the sodium hypoiodite (NaOI) that is formed when I_2 crystals and NaOH solution are added. In turn, the acetaldehyde is halogenated to form triiodoacetaldehyde, which is cleaved by NaOH to form iodoform and sodium formate. The iodoform (CHI_3) appears as yellow crystals with a characteristic medicinal odor.

PROCEDURE

Your instructor will assign you one of the following carbohydrates to set up in accordance with part A: glucose, galactose, fructose, xylose, sucrose, lactose, maltose, and starch. Note whether you are to use an active dry baker's yeast or brewer's yeast.

A. *Fermentation of Sugars in a Smith Fermentation Tube.* (**NOTE:** The amounts of sugar solution and yeast used in the procedure below are based on a fermentation tube having a vertical closed tube with an inside diameter of 13 mm and length of 130 mm. The volume of sugar solution to fill the vertical closed tube and adequately close the constricted bend is about 18 ml. This volume may vary with other sizes of fermentation tubes and may be as high as 30 ml for larger tubes. Approximately 0.25 g yeast is used per 10 ml sugar solution.)

At the beginning of the laboratory period, place 0.5 gram of active dry yeast into the fermentation tube and add 18 ml of 2% sugar solution. Securely close the open end of the tube with your thumb and *completely* suspend the yeast in the solution by tilting the tube back and forth, and by shaking. When the yeast is entirely dispersed, tilt the fermentation tube to permit the solution to flow into the closed end of the tube so that all air is excluded. Set the tube into a labeled 400-ml beaker and place the beaker in a 40° oven. If fermentation occurs in your tube, allow most of the closed tube to fill with gas. The solution will rise toward the open end of the fermentation tube. If no gas forms within 45-60 minutes, leave the tube in the oven and check the fermentation tube occasionally during the remainder of the laboratory period for possible gas formation.

B. ***Test for Carbon Dioxide.*** Carefully remove the beaker and fermentation tube from the oven so that the entrapped gas does not escape. Slowly add 10 ml of 5% NaOH *without* agitation. Now completely fill the remaining space in the open end of the tube with water from a beaker. (Do *not* add water from the faucet, since this would cause premature mixing.) Place your thumb over the open end of the tube and mix contents carefully. Note the great suction produced on your thumb. The instructor will demonstrate the procedure to the entire class. Indicate on your report sheet which carbohydrates fermented and their relative rates of fermentation.

C. ***Test for the Formation of Alcohol.*** In an ordinary test tube place 10 ml of 5% glucose solution, add 0.75 gram of active dry yeast. Set a cork *loosely* on the test tube and set aside until the next laboratory period.

At the next laboratory period *decant* half of the contents of the tube through glass wool in a small funnel. Refilter through a wet filter paper to obtain a clear liquid that can be tested for alcohol by the iodoform test as follows: Add a couple of crystals of iodine and 2 ml of 5% NaOH to the solution. Do not add too much iodine or it will not dissolve. Heat gently in a water bath and note the odor of iodoform. Let stand a few minutes and crystals of iodoform will settle out if enough alcohol was formed from the fermented sugar. Have the instructor check your results.

Reagents and Materials are listed on page 294.

LABORATORY REPORT

Name_____ Section_____ Date_____

Experiment 35. FERMENTATION OF SUGARS

A. *Fermentation of Sugars in a Smith Fermentation Tube*

Type(s) of yeast used._____

Indicate the degree and ease of fermentability of *all* carbohydrates tested by your class.

Glucose_____ Sucrose _____

Galactose_____ Lactose _____

Fructose_____ Maltose _____

Xylose_____ Starch _____

If more than one type of yeast was used by the class, were there any differences in their ability to ferment certain sugars?

_____

_____

What gases are primarily present above the liquid surface

    a. At the *open* end of the tube?_____

    b. At the *closed* end of the tube? _____

What causes the great suction on your thumb after adding NaOH and mixing the contents of the tube?

_____

Write a balanced equation for the chemical reaction occurring between NaOH and the entrapped gas.

_____

B. *Test for Formation of Ethyl Alcohol*

Why is the test tube only *loosely* stoppered?

_____

Give the color and odor of the iodoform crystals.

Color_____ Odor_____

Write a series of equations showing the steps in the formation of iodoform from ethyl alcohol.

QUESTIONS

Name _____ Section _____ Date _____

Experiment 35. FERMENTATION OF SUGARS

1. By means of a flow diagram show the transformation of glucose to pyruvate (give names and structures). At the arrows between each step indicate the enzyme and coenzyme or cofactors, if any.

2. At which steps in the above flow diagram is ATP utilized and generated? What is the net gain in ATP per glucose molecule fermented?

3. Using names and structures show the transformation of pyruvic acid to ethyl alcohol. At the arrows between each step give the names of the enzyme and coenzyme involved.

4. Compared to the relative larger amounts of glucose undergoing fermentation, explain why only small amounts of NAD^+ and NADH are required.

5. Could sucrose, lactose, or maltose be used as a starting material in place of glucose? Give reasons for your answers.

Sucrose_____

Lactose_____

Maltose_____

6. Considering your observations on the fermentability of starch by yeast in this experiment, how do you explain that starch and yeast are materials used in the brewing industry to produce ethyl alcohol? (Review Expt. 32.)

_____

_____

Bioluminescence. Determination of ATP

INTRODUCTION

Many living organisms have the ability to produce visible light, for example, the common firefly, various bacteria, fungi, glowworms, protozoa, fish, shrimps, sponges, and many others. The energy for the light production comes from adenosine-5'-triphosphate, or ATP; chemical energy is converted into light energy by an enzymatically catalyzed process. The nature of the reactions that result in the light production, or *bioluminescence*, has been studied most extensively in the firefly. The luminous organ, or "lantern," has been shown to contain an enzyme, called luciferase, and a special chemical, luciferin, which serves as its substrate. Both are essential for the reaction, as are ATP and magnesium ions. The sequence of events appears to be as follows:

$$(1) \qquad LH_2 + ATP \xrightarrow[Mg^{++}]{\text{luciferase}} LH_2 \cdot AMP + P \cdot P$$

$$\begin{array}{ccc} \text{reduced} & & \text{activated} & \text{inorganic} \\ \text{luciferin} & & \text{luciferin} & \text{pyrophosphate} \end{array}$$

$$(2) \qquad LH_2 \cdot AMP \xrightarrow{O_2} L \cdot AMP + H_2O + \text{visible light}$$

If all other components are present in excess, the light emission is limited by ATP and is directly proportional to the amount of ATP supplied. Other "high energy" phosphorus compounds are ineffective. The system thus provides a means of ATP analysis that is quick, sensitive, and specific. For precise quantitative work the light emitted is measured with a special physical instrument, such as a photofluorometer. In this experiment qualitative estimates will be made by visual observation.

PROCEDURE

Because of the special requirements for organization of this experiment and the necessity to perform it in a semidark area, it is recommended that it be performed as a class project with as much student participation as possible. Although live fireflies may be used (see References), the commercially available desiccated firefly lanterns are more convenient. (See Reagents in Appendix.)

A. *Samples for ATP Analysis.* Muscle or kidney tissue from a freshly-killed animal (rat, chick, or other small animal) is quickly weighed and dropped in boiling water, in the proportion of one part tissue to 10 parts water. Boil for 10-15 minutes, allow to settle and use the supernatant liquid for ATP assay as described in part B. For better extraction of the ATP, it is desirable to cut the tissue into pieces as fine as can be accomplished without undue delay. Fresh baker's yeast samples may be prepared for assay in the same manner.

B. *Estimation of ATP by Bioluminescence.* The experiment should be carried out and the tubes observed in very dim light, after your eyes have become adapted to the dark. A reagent solution consisting of a mixture of 5 ml freshly-prepared firefly lanterns in 0.1 M arsenate buffer, 3.5 ml 0.1 M MgSO₄, and

3.5 ml glycine buffer (pH 7.4) is made immediately before the laboratory period. The experiment is then set up as follows:

1. Pipet 0.7 ml of the above reagent mixture into 10 unlabeled test tubes (13 x 100 mm). Do *NOT* pipet by mouth (arsenate!). Two of the 10 test tubes are extra if needed in step 5.

2. Prepare a standard series of labeled tubes (18 x 150 mm) made up with 0.3 ml ATP solutions of differing concentration. The following concentrations are convenient, but may be varied as desired: 1×10^{-3} M, 5×10^{-4} M, 2×10^{-4} M, 1×10^{-4} M and 5×10^{-5} M. Prepare a sixth tube with 0.3 ml of water in place of ATP solution to serve as a control. All solutions should be at about 23-25°C, as the enzyme is most active at that temperature.

3. Prepare two tubes containing 0.3 ml tissue sample extracts (part A) at two different concentrations in place of using the standard ATP.

4. Into each tube containing the standard ATP, control, and tissue extract, pour 0.7 ml of the firefly solution reagent previously measured in step 1.

5. By visual observation estimate as nearly as possible which ATP tube most closely matches the luminescence of the tissue extract samples. It may be necessary to dilute the extracts to bring them within the range of the standard ATP series. Record whatever information you obtain from your observations.

After the above comparisons have been made, chill one of the brightest tubes in ice for a few minutes and observe, then warm it to 23-25°C once again and note any changes that occur. What do the results tell you about the effect of temperature on the rate of enzyme action? Take another tube that shows good luminescence and add to it 0.2 ml of 5% TCA solution. Place a third in boiling water for a few minutes, then cool and observe.

REFERENCES

CHASE, A. M. 1960. *Methods of biochemical analysis.* ed. David Glick. vol. 8. New York: Interscience Publishers, pp. 62-116.

DeLuca, M. A. Bioluminescence and Chemiluminescence. In *Methods in enzymology.* ed. S. P. Colowick and N. O. Kaplan. Vol. 57. New York: Academic Press.

GREEN, A. A., and McElroy, W. D. 1956. *Biochem. Biophys. Acta* 20:170.

McElroy, W. D. 1955. *Methods in enzymology.* ed. S. P. Colowick and N. O. Kaplan. vol. 2. New York: Academic Press, p. 854.

STREHLER, B. L., and McElroy, W. D. 1957. *Methods in enzymology.* ed. S. P. Colowick and N. O. Kaplan. vol. 3. New York: Academic Press, p. 871.

STREHLER, B. L., and TOTTER, J. R. 1952. *Archiv. Biochem. Biophys.* 40:28.

WHITE, E. H.; McCAPRA, F.; FIELD, G. F.; and McElroy, W. D. 1961. *J. Amer. Chem. Soc.* 83:2402.

Reagents and Materials are listed on page 295.

Name_____ Section_____ Date_____

Experiment 36. BIOLUMINESCENCE. DETERMINATION OF ATP

ATP Standards	Tube No.	Molarity of ATP	Relative Concentrations	Comments
	1	_____	__1.0__	
	2	_____	_____	
	3	_____	_____	
	4	_____	_____	
	5	_____	_____	
Water Control 6		_____	__0.00__	

ATP in Tissue

Kind of tissue used _____

Grams of tissue extracted_____ g   Volume of water used_____ ml

Molarity of standard ATP having same luminescence as sample_____ M

Millimoles ATP per 0.3 ml = (M × 0.3)_____

Calculate millimoles ATP per gram of tissue.

Factors Affecting Luminescence

1. Cooling tube in ice water and rewarming to 23-25°C.

_____

2. Addition of 0.2 ml 5% TCA to a tube.

_____

3. Heating a luminescent tube in boiling water, and cooling.

_____

QUESTIONS

Name _____ Section _____ Date _____

Experiment 36.   BIOLUMINESCENCE. DETERMINATION OF ATP

1. Summarize the evidence you have obtained in this experiment regarding the effect of various chemical and physical agents or conditions on enzyme action.

2. What evidence is there that the light production depends on an enzyme-catalyzed reaction?

   _____

   _____

   _____

3. Was any light produced in the absence of ATP? Explain.

   _____

4. Why did the addition of 5% trichloroacetic acid have an inhibiting effect on the production of luminescence?

   _____

5. List at least 4 different types of energy-requiring processes in the animal body that derive their energy from ATP.

   a. _____

   b. _____

   c. _____

   d. _____

Cytochrome *c* Oxidase. (Cytochrome *aa₃*)

INTRODUCTION

In aerobic organisms the ultimate acceptor of hydrogen removed from various metabolites by the dehydrogenases is oxygen. The enzymes concerned with the final passage of hydrogen to oxygen are called *terminal oxidases*. A powerful thermolabile oxidase that is responsible for 70-80 percent of the oxygen uptake of many aerobic cells can be demonstrated in yeast and animal tissues by appropriate methods. It is called *cytochrome c oxidase*, since it is concerned physiologically with the reoxidation of the reduced cytochromes by oxygen. A reaction that has been used extensively to detect the presence of this enzyme employs the so-called *Nadi* reagent, and results in the formation of a blue-colored indophenol.

The reaction consists in the oxidation of p-aminodimethylaniline by cytochrome *c* (catalyzed by cytochrome *c* oxidase in the presence of oxygen) to the nitroso-compound and the condensation of the latter with α-naphthol to form the colored indophenol, which can be reversibly reduced to a colorless compound.

What once was simply known as cytochrome oxidase is now specifically called *cytochrome c oxidase*, or *cytochrome aa₃*, which is a combination of cytochromes *a* and *a₃* in a single, large respiratory enzyme. This complex of cytochromes *a* and *a₃* contains two heme-iron atoms and two nonheme-copper atoms, which are involved in electron transport with the transitions of $Fe(III) \rightleftarrows Fe(II)$ and $Cu(II) \rightleftarrows Cu(I)$. A substrate (p-aminodimethylaniline in this experiment) passes a pair of electrons to cytochrome *c* and the electrons are transmitted first to cytochrome *a* and then to cytochrome *a₃* and finally to oxygen. It is believed that the copper atoms in the cytochrome *c* oxidase (cytochrome *aa₃*) catalyze the transfer of electrons to oxygen.

Cyanide (HCN or CN^-) and hydrogen sulfide (H_2S) will inhibit the catalytic activity of cytochrome c oxidase. Cyanide strongly binds with metal ions and interferes in the transport of electrons. Hydrogen sulfide evidently binds copper ions, since it is known that H_2S inhibits other copper-containing enzymes.

The Nadi reagent has been used to detect cytochrome c oxidase activity in animal tissues by a histochemical method (Gomori reference).

PROCEDURE

A. *Yeast Suspensions and Nadi Reagent.* A stock suspension of yeast in phosphate buffer (pH 7.3) will be prepared for you. One portion of the stock yeast suspension will be left unheated, and another portion will be heated at 57°C for one hour in order to destroy the activity of reducing systems. Before the start of the laboratory period, the stock yeast suspensions are each diluted ten-fold with phosphate buffer (pH 7.3) to prepare about 50 ml of dilute unheated yeast suspension and 200 ml of the dilute heated yeast suspension (see Appendix).

The Nadi reagent is a *freshly-prepared* mixture of 0.01 M p-aminodimethylaniline, 0.01 M α-napthol, and 0.1% sodium carbonate solutions. The three separate solutions must be mixed immediately before use.

B. *Color Development in Yeast Suspensions.* In each of 5 labeled test tubes, place 2 ml of the appropriate dilute yeast suspension as indicated in the table below. Tube 3 is heated in a *boiling* water bath for 5 minutes and then cooled. Tubes 4 and 5 are treated as indicated. A tube 6 is set up as a color control, using 2 ml of phosphate buffer (pH 7.3) in place of the yeast suspension.

To each of the 6 test tubes is then added 0.5 ml of the freshly-prepared Nadi reagent from the 25-ml buret available on the reagent bench. The tubes are shaken for several minutes and the degree of color development is observed.

Check your results with your instructor and repeat those tubes that show improper color. Tabulate and interpret your results on the report sheet.

Tube No.	Yeast suspension		Additional Treatment	Nadi reagent
	unheated	heated		
	ml	ml		ml
1	2	0	none	0.5
2	0	2	none	0.5
3	0	2	boiled and cooled	0.5
4	0	2	2 − 3 drops 0.01 M KCN	0.5
5	0	2	2 − 3 drops H_2S solution	0.5
6	2 ml phosphate buffer			0.5

REFERENCES

GOMORI, G. 1957. *Methods in enzymology.* ed. S. P. Colowick and N. O. Kaplan. vol. 4. New York: Academic Press, p. 382.

LEHNINGER, A. L. 1975. *Biochemistry.* 2d ed. New York: Worth Publishers, Inc., p. 492.

OSER, B. L., ed. 1965. *Hawk's physiological chemistry.* 14th ed. New York: McGraw-Hill Book Co., pp. 424, 434.

Reagents and Materials are listed on page 296.

Name_____ Section_____ Date_____

Experiment 37.   CYTOCHROME *c* OXIDASE (CYTOCHROME *aa₃*)

Results and Interpretations

Tube No.	Difference in tube contents and treatment from that of tube 2	Color observed
1		
2	— —— —— —— —	
3		
4		
5		
6		

What evidence did you obtain

a. that cytochrome *c* oxidase is a metal-containing enzyme?

_____

b. that the colored indophenol can be reversibly reduced?

_____

c. that the enzyme is thermolabile above 57° C, but not below this temperature?

_____

QUESTIONS

Name_____ Section_____ Date_____

Experiment 37. CYTOCHROME *c* OXIDASE (CYTOCHROME *aa₃*)

1. Outline the preparation of the yeast suspensions (see Appendix) and give reason for the 57° incubation.

_____

_____

_____

2. Why is the Nadi reagent prepared *fresh* before use?

_____

3. Why is cytochrome *c* oxidase also called cytochrome aa_3?

_____

_____

_____

4. Name other cytochromes in the electron transport system (in addition to cytochrome aa_3) and list them in the order in which they transport electrons.

_____

_____

5. Name the prosthetic group of cytochromes and the metal ion associated with this group.

_____

6. What is the function of cytochrome *c* oxidase in yeast?

_____

_____

7. Write a simple, balanced equation showing the acceptance of electrons by O_2 with the formation of water.

_____

Urease. Determination of Urea in Blood

INTRODUCTION

Urease is an enzyme that quantitatively hydrolyzes urea into NH_3 and CO_2. At the pH at which the enzyme acts best, about 7-8, these products combine to form ammonium carbonate, so that the net result of the reaction is:

$$NH_2-CO-NH_2 + 2H_2O \xrightarrow{\text{urease}} (NH_4)_2CO_3$$

The enzyme shows extreme specificity for urea; it attacks no other nitrogenous compounds, not even as closely related a compound as monomethylurea. It is this extreme specificity that makes it possible to use the enzyme directly for quantitative urea estimation in so complex a mixture as blood or urine. The urea-containing sample is treated with urease and the ammonia formed is determined colorimetrically by the Nessler method after removal of interfering proteins by precipitation with sodium tungstate. The reaction of ammonia with a modified Nessler's reagent (a solution of KI and HgI_2) yields a yellow color whose intensity is determined photometrically at 490 nm.

PROCEDURE

A. **Blood Specimens and Urease Solutions.** Oxalated whole blood or blood plasma (2 mg sodium oxalate per ml blood) or blood serum can be used. Heparinized blood can also be used. Fluoride *cannot* be present, because it inhibits urease activity. Commercially available plasma or serum can be used as unknowns.

An *active* urease solution will be prepared for you with an activity of approximately 1.6 Sumner units per ml. A portion of this solution will be boiled to prepare an *inactive* urease solution that will be used in a tube (Enzyme Blank) containing an aliquot of the prepared blood specimen to determine the amount of ammonia-nitrogen initially present in the enzyme and blood. By subtracting this correction from the NH_3-N found in the sample tube (Unknown), you can then calculate the urea-nitrogen in the blood specimen.

NOTE: *Do NOT use acetone to dry pipets or any other glassware.* Acetone (even its vapors) causes precipitation of Nessler's reagent, which is added to develop the color with ammonia.

B. **Hydrolysis of Urea and Precipitation of Proteins.** Pipet 0.5 ml of whole blood, plasma, or serum into each of two clean, dry test tubes labeled "Unknown" and "Enzyme Blank." (Your instructor may give you a prepipetted sample.) All reagent solutions will be provided for you in burets on the reagent bench. Proceed as follows:

1. Add 2.50 ml of water to each tube and mix.
2. Add 0.50 ml of *active* (unboiled) urease solution (0.8 units urease) to the "Unknown" tube and mix.
3. Add 0.50 ml of *inactive* (boiled) urease solution to the "Enzyme Blank" tube and mix.
4. Incubate the tubes in a 35-40° water bath for 15 minutes. (If a laboratory water bath is not available, heat 400 ml water to 45° in a 600-ml beaker, turn off flame, and place the tubes in this bath for 15 min.)
5. After incubation, add 1.00 ml of 10% sodium tungstate to each tube and mix.

6. Add 1.50 ml of 1 N H_2SO_4 to each tube and mix well by swirling.

7. Allow the tubes to stand for 5-10 minutes.

8. Centrifuge as shown by the instructor.

9. Pour off the clear supernatant solution into clean, dry test tubes for the ammonia determination described below. If the supernatant is not clear, filter through a dry filter paper and funnel.

C. **Determination of Ammonia after Nesslerization.** Using four clean, dry test tubes labeled as indicated in step 4, proceed as follows:

1. To each tube add 2.00 ml distilled water.

2. To each tube add 2.00 ml of modified Nessler's reagent from the 50-ml buret provided.

3. To each tube add 3.00 ml 2 N NaOH from the 50-ml buret provided.

4. Add the solution specified below to the appropriately labeled tube:

"Water Blank" .... 2.00 ml distilled water
"Enzyme Blank" .... 2.00 ml *supernatant* from *inactive* enzyme, part B, step 9.
"Unknown" .... 2.00 ml *supernatant* from *active* enzyme, part B, step 9.
"Standard" .... 2.00 ml 0.001 N $(NH_4)_2SO_4$ solution.

5. Thoroughly mix the contents of all tubes and allow them to stand for 20 minutes for color development.

6. Set the Spectronic 20 colorimeter at zero Absorbance (100% Transmittance) by using the "Water Blank" with the wavelength set at 490 nm. Read the color intensities of the "Enzyme Blank," "Unknown," and "Standard" (in that order). Your instructor will review the proper use of the instrument. Read "Notes on the Use of the Spectronic 20" on page 21, and review "Colorimetry" in the introduction to Experiment 4.

7. Record the absorbancies and other data on the report sheet and complete the calculations.

D. **Calculations for Ammonia-Nitrogen.** Since the absorbance (A) value is proportional to the ammonia content (i.e., the color produced by Nesslerization follows Beer's law over the concentration range of 0.01-0.04 mg per tube), the urea content of the unknown sample can now be calculated. First calculate the milligrams of ammonia-N in the standard tube (C_s).

$$C_s = mg\ NH_3\text{-}N/tube = meq\ wt\ N \times ml\ (NH_4)_2SO_4 \times normality\ of\ standard\ (NH_4)_2SO_4$$
$$C_s = 14\ mg/meq \times 2\ ml \times 0.001\ meq/ml$$

Then calculate the mg NH_3-N in "Unknown" tube (C_u)

$$C_u = \frac{net\ A_u}{A_s} \times C_s$$

where A_u = total "Unknown" A minus "Enzyme Blank" A.

The NH_3-N found in the "Unknown" tube is *one-third* (Why?) of that produced by the action of urease on the urea contained in the original 0.5 ml of whole blood, plasma, or serum. Since the NH_3-N comes only from the urea, then the urea-N in 0.5 ml of sample can be now converted to a 100-ml basis as follows:

$$mg\ urea/100\ ml = mg\ urea\text{-}N/100\ ml \times \frac{formula\ wt\ of\ urea}{formula\ wt\ of\ 2\ nitrogen\ atoms}$$

REFERENCE

JOHNSON, M. J. 1941. *J. Biol. Chem.* 137:575.

Reagents, Materials and Equipment are listed on page 297.

LABORATORY REPORT

Name_____ Section_____ Date _____

Experiment 38. UREASE. DETERMINATION OF UREA IN BLOOD

Data and Results

Absorbance (A) values: _____ Total A, unknown blood sample

_____ Enzyme blank

_____ A_u net value, unknown blood sample

_____ A_s, standard tube

Calculations: (Show work and units)

mg ammonia-N in standard tube (C_s)

$C_s =$

_____ mg NH_3-N/tube

mg ammonia-N per unknown sample tube (C_u)

$C_u =$

_____ mg NH_3-N/tube
(4 sig. fig.)

mg urea-N per 0.5 ml unknown sample $=$

_____ mg urea-N/0.5 ml
(4 sig. fig.)

mg urea-N per 100 ml unknown sample $=$

_____ mg urea-N/100 ml
(3 sig. fig.)

Convert urea-N/100 ml to urea/100 ml.

QUESTIONS

Name _____ Section _____ Date _____

Experiment 38.   UREASE. DETERMINATION OF UREA IN BLOOD

1. What is the most important natural source of urease? (See Reagents, p. 297.) _____

   What is the optimum pH range for urease activity? _____

2. How many grams of enzyme-containing powder (2,000 Sumner units per gram) are required to prepare 100 ml of urease solution with an enzyme activity of 4 units per ml? Show work.

3. Write the structures of the following:

   a. Urea   b. Monomethylurea   c. Biuret

4. Considering the similarities in the above structures, discuss the specificity of urease in catalyzing the hydrolysis of amide-type bonds.

   _____

5. In this experiment, what is the purpose of

   a. the enzyme blank? _____

   b. addition of sodium tungstate solution after urea hydrolysis? _____

6. Excess ammonia from amino acid metabolism is converted to urea in the liver.
   a. Give the name and structure of the compound formed at the start of the urea cycle from the reaction of $NH_3 + ATP + CO_2$.

      Name _____   Structure:

   b. Name the amino acid that is enzymatically hydrolyzed in the final step of the cycle to form urea and a regenerated product.

      $$ \underline{\hspace{6cm}} \xrightarrow[\text{enzyme}]{H_2O} \text{urea} + \underline{\hspace{6cm}} $$

7. a. What is the normal range of urea nitrogen in human blood plasma or serum? _____
   b. How many grams of *urea* does this represent in the entire blood volume of an average adult person (assume this to be 5000 ml)?

8. a. About how much urea is excreted daily by a human adult? _____
   b. How is this related to the diet?

   _____

Nucleic Acids

Experiment 39

Preparation and Properties of Nucleic Acids

INTRODUCTION

Ribonucleic acids (RNA) and deoxyribonucleic acids (DNA) of many kinds occur in all living cells. They are moderate to very large-sized polymers composed of mononucleotide units joined by phosphate ester linkages. Separation of the higher molecular weight nucleic acids from other tissue components without partial decomposition is very difficult, but preparations suitable for study of chemical and physical properties can be readily obtained from many tissues. If it is not possible to perform the first part of this experiment because liver tissue from rats is not available, the color tests for nucleic acid components can be carried out on commercially available RNA and DNA materials. In either case, the student should be familiar with a successful method of separating nucleic acids from other tissue components.

In the separation procedure, the tissue is homogenized in cold trichloroacetic acid (TCA) solution to break the cells and release the nucleic acids in an insoluble form. After centrifugation and removal of soluble materials, the pellet (precipitate) is treated with hot TCA solution that dissolves nucleic acids but not proteins. The material that remains soluble is an impure mixture consisting mainly of partially degraded nucleic acids.

Complete hydrolysis of a nucleic acid gives a mixture of 4 constituent bases, a pentose sugar, and one mole of phosphoric acid per mole of base. Various chemical conditions as well as specific enzymes have been employed for partial hydrolysis to particular fragments, such as nucleotides, nucleosides, ribose phosphates, and free bases. Review the nature of these products in your text or reference book. The color tests on RNA and DNA given below are carried out under hot, acidic conditions, which will effect partial or complete hydrolysis of the nucleic acids.

PROCEDURE

A. *Homogenization and Extraction of Rat Liver.* Livers from freshly-killed rats are packed in ice until thoroughly chilled. Homogenization is done in a Waring blender by grinding the tissue 30-60 seconds with 4 parts (by weight) of ice-cold distilled water to give a uniform suspension. Prepare immediately before use and store in ice.

To 2 ml of a 20% rat liver homogenate add 5 ml of cold 2.5% trichloroacetic acid (TCA) solution and mix. (**CAUTION:** Do not pipet TCA by mouth.) Centrifuge the mixture at top speed for 2 minutes in a clinical centrifuge, using a 15-ml conical centrifuge tube. Discard the supernatant fluid, resuspend the pellet (sedimented solid) in 2.5 ml of the TCA solution by carefully grinding with a 3-mm diameter glass rod, and recentrifuge. Wash the pellet twice with 5-ml portions of 95% ethanol, resuspending and centrifuging each time as before. Finally resuspend the remaining pellet in 1.3 ml of water and then add 1.3 ml of 10% TCA and heat for 15 minutes at 90°C (thermometer in liquid) with occasional agitation. Centrifuge, decant the supernatant liquid and save it. Resuspend the pellet in 2.5 ml of 5% TCA, recentrifuge, and add the liquid to the previous supernatant. The combined solution (recentrifuged if not clear) contains nucleic acids from the original tissue that were first precipitated by the cold TCA and then solubilized by heating in TCA solution. Use portions of the solution for the tests in part B.

B. **Color Tests for Nucleic Acid Components.** If the preparation of the nucleic acid solution in part A was not carried out, you will obviously omit it from the color tests given below.

(1) *Ribose.* In separate test tubes perform the tests individually on 0.5 ml of the following solutions:

(a) The nucleic acid solution prepared in part A.

(b) 0.02% solution of commercial ribonucleic acid (RNA) in 5% TCA.

(c) 0.05% solution of commercial deoxyribonucleic acid (DNA) in 5% TCA.

(d) 0.01% solution of ribose or xylose (an aldopentose).

(e) 0.01% solution of glucose (an aldohexose).

To each tube add 2 ml of orcinol reagent from the 50-ml buret on the reagent bench. (**CAUTION:** Contains conc. HCl. Do not pipet by mouth.) Mix and place the tubes in a boiling water bath for 5-10 minutes. Describe and report the color change that occurred in each case. What are your conclusions? (See Expt. 11, p. 77.)

(2) *Deoxyribose.* In separate test tubes perform the test individually on 1 ml of the same solutions used in (1) above. To each test tube add 2 ml of diphenylamine reagent from the 50-ml buret on the reagent bench. (**CAUTION:** Contains glacial acetic acid and conc. H_2SO_4. Do not pipet by mouth; use the buret provided.) Mix and heat the tubes in a boiling water bath for 10 minutes. Record your observations.

(3) *Phosphate.* To 1 ml of your nucleic acid solution from part A (or 1 ml of the 0.02% commercial RNA solution) add 1 ml (20 drops) of 5 N H_2SO_4 and heat the mixture over a *small* flame, shaking frequently until the contents of the tube become brown (use test tube holder). Cool, add 1 drop of 2 N HNO_3 and heat until white fumes just appear. If liquid is not colorless at this point, repeat the addition of a drop of HNO_3 and reheat. When the liquid is colorless, cool, add 1 ml of water, heat 5 minutes in a boiling water bath, and cool again. Add 1 ml (20 drops) of molybdate reagent and, after mixing, 1 ml of reducing reagent (p-methylaminophenòl (Elon) in $NaHSO_3$ solution). Mix well, dilute to 9-10 ml with water, let stand 10 minutes and observe the color. Compare the result with the color produced when the test is applied to 1 ml of 0.001 M NaH_2PO_4 solution. (NOTE: The phosphate combines with molybdate and the hexavalent molybdenum of the phosphomolybdate is reduced by the Elon and bisulfite mixture to a blue complex reduction product.)

(4) *Purine and pyrimidine bases.* Place a little guanine in a small evaporating dish, add a few drops of concentrated nitric acid, and evaporate to dryness very carefully on a water bath. The yellow residue upon moistening with 10% KOH becomes red in color and upon further heating assumes a purplish red hue. Now add a few drops of water and warm. A yellow solution results, which yields a red residue upon evaporation. Apply the test also to adenine and to uric acid. This is called the murexide test.

Treat 2-3 ml of a saturated aqueous solution of cytosine or uracil with an excess of bromine water, adding it dropwise with shaking until the solution is colored. Remove the excess of bromine by gently boiling the solution. Add a slight excess of barium hydroxide, about 0.1 g on tip of scoopula. (Keep bottle tightly closed when not in use.) After mixing, a purple color indicates cytosine or uracil and is due to the purple barium salt of dialuric acid (2,4,6-oxy-5-hydroxypyrimidine). The test is negative for thymine.

REFERENCE

SCHNEIDER, W. C. 1972. In *Methods in enzymology.* vol. 3. ed S. P. Colowick and N. O. Kaplan. New York: Academic Press, p. 680.

Reagents, Materials, and Equipment are listed on page 297.

LABORATORY REPORT

Name_____ Section_____ Date_____

Experiment 39. PREPARATION AND PROPERTIES OF NUCLEIC ACIDS

Color Tests for Nucleic Acid Components

1. *Ribose*—Describe the color changes with the orcinol reagent. What are your conclusions?

   Nucleic acid preparation _____

   0.01% aldopentose solution _____

   0.01% aldohexose solution _____

   0.02% commercial RNA solution _____

   0.05% commercial DNA solution _____

2. *Deoxyribose*—Describe the color changes with diphenylamine reagent. What are your conclusions?

   Nucleic acid preparation _____

   0.01% aldopentose solution _____

   0.01% aldohexose solution _____

   0.02% commercial RNA solution _____

   0.05% commercial DNA solution _____

3. *Phosphate*—Describe the color with molybdate reagent. What are your conclusions?
   Nucleic acid preparation or
   0.02% commercial RNA solution. _____

   0.001 M phosphate solution. _____

4. *Purine and Pyrimidine Bases*—Describe the colors formed in the murexide test for guanine, adenine, and uric acid

   _____

   _____

   and in the dialuric acid test for cytosine or uracil.

   _____

   _____

Name_____ Section _____ Date _____

Experiment 39.   PREPARATION AND PROPERTIES OF NUCLEIC ACIDS

1. What is the chief difference between liver homogenate and chopped or ground liver?

_____

_____

Why is the homogenate needed for this experiment?

_____

2. What major chemical constituents and about what percentage of them are present in fresh liver? Consult a general reference book, such as Biochemists' Handbook.

_____

_____

3. Which of the materials in (2) are precipitated by cold TCA?

_____

What type of material remained insoluble after heating in the 5% TCA?

_____

4. From the results of this experiment construct a flow diagram showing how nucleic acids can be separated from the other major constituents of animal tissue.

5. Which color test is given by RNA that is negative with DNA, and vice versa? Would it be possible to test for the presence of either type in the presence of the other?

_____

_____

6. Aside from the sugar components, what other chemical differences exist between RNA and DNA?

_____

_____

7. Which of the compounds tested in part B (4) were purines; pyrimidines?

Purine bases: _____

Pyrimidine bases: _____

8. Write out the ring system for a purine; a pyrimidine.

Purine ring: Pyrimidine ring:

9. Which type of RNA is present in cell cytoplasm

in the highest percent? _____

in the lowest percent? _____

10. Which type of RNA has the

lowest molecular weight? _____

highest molecular weight? _____

General References

ADAMS, C. F. 1975. *Nutritive value of American foods.* U.S. Dept. of Agr. Handbook No. 456. Washington, D.C.: Supt. of Documents, U.S. Government Printing Office.

CLARK, JOHN M., JR., and SWITZER, ROBERT L. 1977. *Experimental biochemistry.* 2d ed. San Francisco: W. H. Freeman and Co.

CONN, E. E., and STUMPF, P. K. 1976. *Outlines of biochemistry.* 4th ed. New York: John Wiley & Sons.

FASMAN, GERALD D., ed. 1975-77. *Handbook of biochemistry and molecular biology.* 3rd ed. Cleveland: CRC Press, 5 v. in 8.

HAUROWITZ, FELIX. 1963. *The chemistry and function of proteins.* 2d ed. New York: Academic Press.

HOLUM, J. R. 1978. *Fundamentals of general, organic, and biological chemistry.* New York: John Wiley & Sons.

JAMES, M. L.; SCHRECK, J. O.; and BeMILLER, J. N. 1980. *General, organic, and biological chemistry.* Lexington, Mass.: D. C. Heath and Co.

LEHNINGER, A. L. 1975. *Biochemistry.* 2d ed. New York: Worth Publishers, Inc.

OSER, B. L., ed. 1965. *Hawk's physiological chemistry.* 14th ed. New York: McGraw-Hill Book Co.

PATAKI, G. 1969. *Techniques of thin-layer chromatography.* Rev. 2d English ed. Ann Arbor, Mich.: Science Publications, Inc.

SEGEL, I. H. 1976. *Biochemical calculations.* 2d ed. New York: John Wiley & Sons.

SMITH, IVOR. 1976. *Chomatographic and electrophoretic techniques,* vol. 1. 4th ed. Chicago: Year Book Medical Pub.

STRYER, L. 1981. *Biochemistry.* 2d ed. San Francisco: W. H. Freman and Co.

SUTTIE, J. W. 1977. *Introduction to biochemistry.* 2d ed. New York: Holt, Rinehart & Winston.

UMBREIT, W. W.; BURRIS, R. H.; and STAUFFER, J. F. 1972. *Manometric techniques.* 5th ed. Minneapolis, Minn.: Burgess Publishing Co.

WATT, B. K., and MERRILL, A. L. 1963. *Composition of foods.* U.S. Dept. of Agr. Handbook No. 8. Washington, D.C.: Supt. of Documents, U.S. Government Printing Office.

WHISTLER, R. L., and WOLFROM, M. L. 1962. *Methods in carbohydrate chemistry.* Analysis and preparation of sugars, Vol. 1. New York: Academic Press.

WINDHOLZ, MARTHA, ed. 1976. *The Merck index of chemicals and drugs.* 9th ed. Rahway, N. J.: Merck & Co., Inc.

ZWEIG, G., and WHITAKER, J. R. 1967-70. *Paper chromatography and electrophoresis.* New York: Academic Press.

Appendix

Atomic Weights

Element	Symbol	Atomic number	Atomic weight	Element	Symbol	Atomic number	Atomic weight
Actinium	Ac	89	(227)	Manganese	Mn	25	54.9380
Aluminum	Al	13	26.98154	Mendelevium	Md	101	(258)
Americium	Am	95	(243)	Mercury	Hg	80	200.59
Antimony	Sb	51	121.75	Molybdenum	Mo	42	95.94
Argon	Ar	18	39.948	Neodymium	Nd	60	144.24
Arsenic	As	33	74.9216	Neon	Ne	10	20.179
Astatine	At	85	(210)	Neptunium	Np	93	237.0482*
Barium	Ba	56	137.34	Nickel	Ni	28	58.70
Berkelium	Bk	97	(247)	Niobium	Nb	41	92.9064
Beryllium	Be	4	9.01218	Nitrogen	N	7	14.0067
Bismuth	Bi	83	208.9804	Nobelium	No	102	(255)
Boron	B	5	10.81	Osmium	Os	76	190.2
Bromine	Br	35	79.904	Oxygen	O	8	15.9994.
Cadmium	Cd	48	112.40	Palladium	Pd	46	106.4
Calcium	Ca	20	40.08	Phosphorus	P	15	30.97376
Californium	Cf	98	(251)	Platinum	Pt	78	195.09
Carbon	C	6	12.011	Plutonium	Pu	94	(244)
Cerium	Ce	58	140.12	Polonium	Po	84	(209)
Cesium	Cs	55	132.9054	Potassium	K	19	39.098
Chlorine	Cl	17	35.453	Praseodymium	Pr	59	140.9077
Chromium	Cr	24	51.996	Promethium	Pm	61	(145)
Cobalt	Co	27	58.9332	Protactinium	Pa	91	231.0359*
Copper	Cu	29	63.546	Radium	Ra	88	226.0254*
Curium	Cm	96	(247)	Radon	Rn	86	(222)
Dysprosium	Dy	66	162.50	Rhenium	Re	75	186.207
Einsteinium	Es	99	(254)	Rhodium	Rh	45	102.9055
Element 104		104	(261)	Rubidium	Rb	37	85.4678
Element 105		105	(262)	Ruthenium	Ru	44	101.07
Element 106		106	(263)	Samarium	Sm	62	150.4
Erbium	Er	68	167.26	Scandium	Sc	21	44.9559
Europium	Eu	63	151.96	Selenium	Se	34	78.96
Fermium	Fm	100	(257)	Silicon	Si	14	28.086
Fluorine	F	9	18.99840	Silver	Ag	47	107.868
Francium	Fr	87	(223)	Sodium	Na	11	22.98977
Gadolinium	Gd	64	157.25	Strontium	Sr	38	87.62
Gallium	Ga	31	69.72	Sulfur	S	16	32.06
Germanium	Ge	32	72.59	Tantalum	Ta	73	180.9479
Gold	Au	79	196.9665	Technetium	Tc	43	(97)
Hafnium	Hf	72	178.49	Tellurium	Te	52	127.60
Helium	He	2	4.00260	Terbium	Tb	65	158.9254
Holmium	Ho	67	164.9304	Thallium	Tl	81	204.37
Hydrogen	H	1	1.0079	Thorium	Th	90	232.0381*
Indium	In	49	114.82	Thulium	Tm	69	168.9342
Iodine	I	53	126.9045	Tin	Sn	50	118.69
Iridium	Ir	77	192.22	Titanium	Ti	22	47.90
Iron	Fe	26	55.847	Tungsten	W	74	183.85
Krypton	Kr	36	83.80	Uranium	U	92	238.029
Lanthanum	La	57	138.9055	Vanadium	V	23	50.9414
Lawrencium	Lr	103	(260)	Xenon	Xe	54	131.30
Lead	Pb	82	207.2	Ytterbium	Yb	70	173.04
Lithium	Li	3	6.941	Yttrium	Y	39	88.9059
Lutetium	Lu	71	174.97	Zinc	Zn	30	65.38
Magnesium	Mg	12	24.305	Zirconium	Zr	40	91.22

Based on 1973 IUPAC Atomic Weights of the Elements.
A value given in parentheses denotes the mass number of the longest-lived isotope.
*Atomic weight of most commonly available long-lived isotope.

Four Place Logarithms

N	0	1	2	3	4	5	6	7	8	9	N	0	1	2	3	4	5	6	7	8	9
10	0000	0043	0086	0128	0170	0212	0253	0294	0334	0374	55	7404	7412	7419	7427	7435	7443	7451	7459	7466	7474
11	0414	0453	0492	0531	0569	0607	0645	0682	0719	0755	56	7482	7490	7497	7505	7513	7520	7528	7536	7543	7551
12	0792	0828	0864	0899	0934	0969	1004	1038	1072	1106	57	7559	7566	7574	7582	7589	7597	7604	7612	7619	7627
13	1139	1173	1206	1239	1271	1303	1335	1367	1399	1430	58	7634	7642	7649	7657	7664	7672	7679	7686	7694	7701
14	1461	1492	1523	1553	1584	1614	1644	1673	1703	1732	59	7709	7716	7723	7731	7738	7745	7752	7760	7767	7774
15	1761	1790	1818	1847	1875	1903	1931	1959	1987	2014	60	7782	7789	7796	7803	7810	7818	7825	7832	7839	7846
16	2041	2068	2095	2122	2148	2175	2201	2227	2253	2279	61	7853	7860	7868	7875	7882	7889	7896	7903	7910	7917
17	2304	2330	2355	2380	2405	2430	2455	2480	2504	2529	62	7924	7931	7938	7945	7952	7959	7966	7973	7980	7987
18	2553	2577	2601	2625	2648	2672	2695	2718	2742	2765	63	7993	8000	8007	8014	8021	8028	8035	8041	8048	8055
19	2788	2810	2833	2856	2878	2900	2923	2945	2967	2989	64	8062	8069	8075	8082	8089	8096	8102	8109	8116	8122
20	3010	3032	3054	3075	3096	3118	3139	3160	3181	3201	65	8129	8136	8142	8149	8156	8162	8169	8176	8182	8189
21	3222	3243	3263	3284	3304	3324	3345	3365	3385	3404	66	8195	8202	8209	8215	8222	8228	8235	8241	8248	8254
22	3424	3444	3464	3483	3502	3522	3541	3560	3579	3598	67	8261	8267	8274	8280	8287	8293	8299	8306	8312	8319
23	3617	3636	3655	3674	3692	3711	3729	3747	3766	3784	68	8325	8331	8338	8344	8351	8357	8363	8370	8376	8382
24	3802	3820	3838	3856	3874	3892	3909	3927	3945	3962	69	8388	8395	8401	8407	8414	8420	8426	8432	8439	8445
25	3979	3997	4014	4031	4048	4065	4082	4099	4116	4133	70	8451	8457	8463	8470	8476	8482	8488	8494	8500	8506
26	4150	4166	4183	4200	4216	4232	4249	4265	4281	4298	71	8513	8519	8525	8531	8537	8543	8549	8555	8561	8567
27	4314	4330	4346	4362	4378	4393	4409	4425	4440	4456	72	8573	8579	8585	8591	8597	8603	8609	8615	8621	8627
28	4472	4487	4502	4518	4533	4548	4564	4579	4594	4609	73	8633	8639	8645	8651	8657	8663	8669	8675	8681	8686
29	4624	4639	4654	4669	4683	4698	4713	4728	4742	4757	74	8692	8698	8704	8710	8716	8722	8727	8733	8739	8745
30	4771	4786	4800	4814	4829	4843	4857	4871	4886	4900	75	8751	8756	8762	8768	8774	8779	8785	8791	8797	8802
31	4914	4928	4942	4955	4969	4983	4997	5011	5024	5038	76	8808	8814	8820	8825	8831	8837	8842	8848	8854	8859
32	5051	5065	5079	5092	5105	5119	5132	5145	5159	5172	77	8865	8871	8876	8882	8887	8893	8899	8904	8910	8915
33	5185	5198	5211	5224	5237	5250	5263	5276	5289	5302	78	8921	8927	8932	8938	8943	8949	8954	8960	8965	8971
34	5315	5328	5340	5353	5366	5378	5391	5403	5416	5428	79	8976	8982	8987	8993	8998	9004	9009	9015	9020	9025
35	5441	5453	5465	5478	5490	5502	5514	5527	5539	5551	80	9031	9036	9042	9047	9053	9058	9063	9069	9074	9079
36	5563	5575	5587	5599	5611	5623	5635	5647	5658	5670	81	9085	9090	9096	9101	9106	9112	9117	9122	9128	9133
37	5682	5694	5705	5717	5729	5740	5752	5763	5775	5786	82	9138	9143	9149	9154	9159	9165	9170	9175	9180	9186
38	5798	5809	5821	5832	5843	5855	5866	5877	5888	5899	83	9191	9196	9201	9206	9212	9217	9222	9227	9232	9238
39	5911	5922	5933	5944	5955	5966	5977	5988	5999	6010	84	9243	9248	9253	9258	9263	9269	9274	9279	9284	9289
40	6021	6031	6042	6053	6064	6075	6085	6096	6107	6117	85	9294	9299	9304	9309	9315	9320	9325	9330	9335	9340
41	6128	6138	6149	6160	6170	6180	6191	6201	6212	6222	86	9345	9350	9355	9360	9365	9370	9375	9380	9385	9390
42	6232	6243	6253	6263	6274	6284	6294	6304	6314	6325	87	9395	9400	9405	9410	9415	9420	9425	9430	9435	9440
43	6335	6345	6355	6365	6375	6385	6395	6405	6415	6425	88	9445	9450	9455	9460	9465	9469	9474	9479	9484	9489
44	6435	6444	6454	6464	6474	6484	6493	6503	6513	6522	89	9494	9499	9504	9509	9513	9518	9523	9528	9533	9538
45	6532	6542	6551	6561	6571	6580	6590	6599	6609	6618	90	9542	9547	9552	9557	9562	9566	9571	9576	9581	9586
46	6628	6637	6646	6656	6665	6675	6684	6693	6702	6712	91	9590	9595	9600	9605	9609	9614	9619	9624	9628	9633
47	6721	6730	6739	6749	6758	6767	6776	6785	6794	6803	92	9638	9643	9647	9652	9657	9661	9666	9671	9675	9680
48	6812	6821	6830	6839	6848	6857	6866	6875	6884	6893	93	9685	9689	9694	9699	9703	9708	9713	9717	9722	9727
49	6902	6911	6920	6928	6937	6946	6955	6964	6972	6981	94	9731	9736	9741	9745	9750	9754	9759	9763	9768	9773
50	6990	6998	7007	7016	7024	7033	7042	7050	7059	7067	95	9777	9782	9786	9791	9795	9800	9805	9809	9814	9818
51	7076	7084	7093	7101	7110	7118	7126	7135	7143	7152	96	9823	9827	9832	9836	9841	9845	9850	9854	9859	9863
52	7160	7168	7177	7185	7193	7202	7210	7218	7226	7235	97	9868	9872	9877	9881	9886	9890	9894	9899	9903	9908
53	7243	7251	7259	7267	7275	7284	7292	7300	7308	7316	98	9912	9917	9921	9926	9930	9934	9939	9943	9948	9952
54	7324	7332	7340	7348	7356	7364	7372	7380	7388	7396	99	9956	9961	9965	9969	9974	9978	9983	9987	9991	9996

Locker Equipment

_____   Lab. Section No. _____ Desk No. _____

Student's Full Name (Please Print)

(NOTE TO INSTRUCTOR: This is a suggested list that may be used at the beginning and end of the semester for checking equipment in student lockers.)

CHECK-IN: Examine all equipment and check it against this list. Replace missing or unsatisfactory items. Then have instructor check your desk and sign your equipment list. Instructor will keep completed list. YOU WILL BE CHARGED FOR LOST KEY OR LOCK.

CHECK-OUT: Return excess equipment to Store. Replace missing or unsatisfactory articles. Don't throw away repairable glassware. Clean rubber stoppers and put them in box provided. Throw dirty rags in waste container. Put new paper in desk drawers. Clean all equipment. Then have instructor check your desk and sign your equipment list. Instructor will keep completed list and desk key.

(Checked in) _____ (Checked out) _____

Instructor's Signature   Instructor's Signature

GLASSWARE:

2 Beakers,   50 ml
2 Beakers,   100 ml
2 Beakers,   150 ml
2 Beakers,   250 ml
2 Beakers,   400 ml
1 Beaker,   600 ml
1 Beaker,   800 ml
1 Beaker,   1000 ml
2 Burets, stopcock, 25 ml and 50 ml
1 Cylinder, graduated, 10 ml
1 Cylinder, graduated, 25 ml, w/guard
1 Cylinder, graduated, 100 ml w/guard
2 Flasks, Erlenmeyer,   50 ml
2 Flasks, Erlenmeyer, 125 ml
2 Flasks, Erlenmeyer, 250 ml
1 Flask, volumetric,   50 ml
2 Funnels, long stem, 65 mm
1 Pipet, volumetric,   1 ml
1 Pipet, volumetric,   5 ml
1 Pipet, volumetric, 10 ml
1 Pipet, graduated,   1 ml
1 Pipet, graduated, 10 ml
1 Thermometer, 110°C
16 Test tubes, Pyrex, 18 x 150 mm, lipless
2 Watch glasses, 75 mm

METALWARE:

2 Bunsen burners
2 Wire gauzes
1 Holder, buret, double
1 Scoopula (Fisher Scientific Co.)
1 Spatula, nickel, double end
1 Tong, crucible
1 Wing top

MISCELLANEOUS:

1 Bottle, wash, polyethylene
1 Bottle, polyethylene, 500 ml
1 Box of matches
1 Brush, test tube, 18 mm
2 Bulbs, rubber suction, small and large
3 Medicine droppers, 2 ml
2 Dishes, porcelain evaporating, 75 mm
1 Spot plate, porcelain, 12 depressions
1 File, triangular, 4 inch
1 Pencil, red, glass marking
1 Sponge
1 Wooden test tube block
1 Wooden test tube holder, pinch type
1 Wooden support, funnel, 4 hole

Common Acids and Bases

Concentrated Solutions of Acids and Bases

	Formula Weight	Grams solute per liter	Specific gravity	Percent by weight	Approximate molarity
Acetic acid, glacial	60.05	1047	1.05	99.7	17.4
Hyrdochloric acid	36.46	433	1.185	36.5	11.9
Nitric acid	63.01	994	1.42	70	15.8
Sulfuric acid	98.08	1766	1.84	96	18.0
Ammonium hydroxide	17.0 (NH_3)	252	0.90	28	14.8
Potassium hydroxide	56.11	757	1.51	50	13.5
Sodium hydroxide	40.00	763	1.53	50	19.1

Dispensing Burets for Concentrated Acids

Special dispensing burets which can be attached to the commercial 5-pint acid bottles are strongly recommended for safety, convenience, and the avoidance of waste. They are available from SGA Scientific, Inc., 737 Broad St., Bloomfield, New Jersey, 07003. The catalog numbers are JB-7552, Buret, automatic, acid, with Teflon stopcock, and JB-7570, Buret adapter (polyethylene) or JB-7571, Buret adapter (Teflon). The following buret volumes are suggested: 25-ml buret for H_2SO_4, 50-ml buret for HNO_3, and 100-ml buret for HCl.

Dilute Solutions of Acids and Bases

The table below indicates the volume of concentrated acid which is *slowly* added to water with *constant mixing* and is diluted to 1 liter to give the molarity/normality specified. Cooling is often necessary when preparing dilute sulfuric acid solutions.

Acid	Molarity or Normality	Milliliters conc. acid	Acid	Molarity or Normality	Milliliters conc. acid
Acetic	1.0	57.5	Nitric	3.0	190
	0.1	5.75		1.0	63.3
				0.1	6.35
Hydrochloric	6.0	504	Sulfuric	2.5 M, 5.0 N	139
	3.0	252		1.0 M, 2.0 N	55.6
	1.0	84		0.5 M, 1.0 N	27.8
	0.1	8.4		0.05 M, 0.1 N	2.8

In general, 0.1 M (0.1 N) acetic, hydrochloric, and nitric acids are readily prepared by diluting 100 ml of 1.0 M (1.0 N) of the acid to 1 liter. For sulfuric acid, dilution of 100 ml of 0.5 M (1 N) H_2SO_4 to 1 liter yields a 0.05 M (0.1 N) H_2SO_4.

Ammonium hydroxide (Ammonia)

1.0 M (1 N)   Dilute 67.5 ml conc. ammonium hydroxide to 1 liter. Store in a tightly stoppered bottle.
0.1 M (0.1 N)   Dilute 100 ml 1 M NH_4OH to 1 liter. Store in a tightly stoppered bottle.

Sodium or Potassium Hydroxides

The table below indicates the grams of sodium hydroxide and potassium hydroxide to prepare 1 liter of the molarity specified. When preparing these aqueous solutions, *constant mixing is necessary* (usually with cooling) until the alkali is completely dissolved.

	Grams per liter	
	NaOH	*KOH*
1.0 M	40.0	56.1
0.1 M	4.0	5.6

Preparation and Standardization of 0.1 N and 1 N Solutions

Commercial Standard Solutions

1   Normal solutions can be accurately prepared from commercially available concentrates of CH_3COOH, HCl, H_2SO_4, and NaOH. When diluted to exactly one liter in a volumetric flask, the normality is certified to be 1.000 N. (J. T. Baker Chemicals, Fisher Scientific, etc.)

Exactly 0.100 N solutions can be prepared by pipetting precisely 100 ml of a 1.000 N solution into a 1-liter volumetric flask and diluting to 1 liter. **CAUTION:** Do NOT pipet by mouth, use a pipet filler (e.g., Propipette).

Laboratory-Prepared Standard Solutions

(1) *Sodium hydroxide*   An approximately 1 N or 0.1 N solution of NaOH can be prepared by dissolving 40 g or 4.0 g, respectively, of reagent grade NaOH pellets (low in carbonate) in water to make 1 liter. Alternatively, 1 N or 0.1 N NaOH solutions can be prepared by diluting 53 ml or 5.3 ml, respectively, of 50% (w/w) NaOH solution (from which the Na_2CO_3 has settled out) to 1 liter. Avoid disturbing the carbonate sediment when measuring the alkali. The NaOH solutions are then standardized as described below. It is better, however, to first prepare and standardize a 1.0 N NaOH solution, and then dilute portions of this standardized solution to 0.100 N NaOH as needed.

(2) *Sulfuric acid and hydrochloric acid*   Approximately 1 N and 0.1 N solutions of H_2SO_4 or HCl can be prepared by diluting to 1 liter the volumes of concentrated H_2SO_4 or HCl specified above in the table under "Dilute Solutions of Acids and Bases." It is preferable to prepare and standardize 1 N solutions of H_2SO_4 or HCl, and then dilute portions to 0.100 N solutions as needed.

Standardization of Alkali or Acid Solutions

Before standardization, the 1 N solutions should be diluted by pipetting 10.0 ml (use pipet filler) of the 1 N solution into a 100-ml volumetric flask and diluting to the mark.

(1) *0.1 N or 1 N NaOH solutions*   Accurately weigh several samples of 0.4 to 0.6 grams of primary standard potassium biphthalate and quantitatively transfer them to 250-ml Erlenmeyer flasks. Dissolve in about 50 ml of distilled water and add 3 drops of 1% phenolphthalein indicator. Titrate the biphthalate with the 0.1 N NaOH, or the diluted 1 N NaOH, until a faint pink color persists. The calculated normality can be used, or the solution can be adjusted to exactly 0.100 N or 1.000 N, if desired.

If standard 0.1 N H_2SO_4 or 0.1 N HCl are available, the NaOH solutions can be standardized against 25.0 ml of either acid in place of the biphthalate.

(2) *0.1 N or 1 N acids*   Into 250-ml Erlenmeyer flasks, pipet several 25.0 ml aliquots of the 0.1 N, or diluted 1 N, acid to be standardized. Add about 25 ml of water and 3 drops of 1% phenolphthalein indicator. Titrate with standard 0.1 N NaOH until a faint pink color persists. Calculate the normality of the acid and adjust the solution to exactly 0.100 N, or 1.000 N, if desired.

NOTE: Solutions of 0.1 M acetic acid and 0.1 M ammonium hydroxide to be used in Experiment 7 (Buffers) should be prepared from stock 1 M solutions after the 1 M solutions have been standardized. The 1 M acetic acid is diluted ten-fold and titrated as described in (2) above. The 1 M NH_4OH is diluted tenfold and titrated using methylene blue-methyl red indicator as described for household ammonia in Experiment 8, page 59.

Reagents, Materials and Equipment

The amounts of reagents and materials indicated in the left column are, in most cases, the *actual* amounts used *per 100 students* if the students use the quantities specified in each experiment. The instructor and stockroom personnel can multiply by the appropriate factor to compensate for smaller or larger classes. In addition, a discretionary allowance should be made for waste or spillage by students. In many instances, larger amounts of *stable* solutions may be prepared for future use. Those materials which are on the locker equipment list (p. 267) or are readily available in most laboratories, are usually omitted in the list given below.

Use distilled water in preparing all aqueous solutions. All chemicals should be of reagent grade.

Experiment 1.   Occurrence of Water in Biological Matter

200	*Aluminum foil weighing dishes,* 60 mm dia.
3-6	*Balances.* Triple-beam, dial type, or top-loading.
20	*Desiccators,* 200 mm i.d., containing anhydrous $CaCl_2$ or anhydrous $CaSO_4$.
25 g	*Cupric sulfate,* $CuSO_4$, anhydrous white powder. If the white powder acquires a bluish tinge after frequent openings of reagent bottle, dehydrate the reagent by heating carefully in a porcelain dish at a temperature not to exceed 275°C (electric furnace).
— —	*Samples:* egg white, fat (lard), meat, potato.

Experiment 2.   Dialysis

125 ft.	*Dialyzing tubing,* seamless, reconstituted cellulose, 33 mm (1.3 inches) flat width.
3-6	*Funnel,* short stem. (Stockroom or reagent shelf)
5 liters	*Milk,* whole or skim.
100 ml	*Acetic acid,* 6 M. Dilute 86 ml glacial acetic acid to 250 ml with water. Dispense from a dropper bottle.
1 liter	*Ammonium chloride,* 10% solution. Dissolve 100 g NH_4Cl in water and make to 1 liter.
100 ml	*Ammonium hydroxide,* 2 M. Dilute 67 ml of conc. NH_4OH to 500 ml. Dispense from a dropper bottle.
1.5 liters	*Ammonium molybdate solution.* Dissolve 100 grams of molybdic acid in 144 ml of conc. ammonium hydroxide (sp. gr. 0.90) and 271 ml of water; slowly and with constant stirring pour the solution thus obtained into 489 ml of conc. nitric acid (sp. gr. 1.42) and 1148 ml of water. Keep the mixture in a warm place for several days, or until a portion heated to 40°C deposits no yellow precipitate of ammonium phosphomolybdate. Decant the solution from any sediment and preserve in glass-stoppered bottles.
500 ml	*Ammonium oxalate,* saturated solution. Dissolve 50 g of $(NH_4)_2C_2O_4 \cdot H_2O$ in water with warming, and allow to cool.
1 liter	*Benedict's reagent.* See sugar reagents for Expt. 10, page 276.
150 ml	*Copper sulfate,* 0.1% solution. Dissolve 0.2 g anhydrous $CuSO_4$ or 0.3 g $CuSO_4 \cdot 5H_2O$ in 200 ml water. Dispense from a dropper bottle.
— —	*Methyl red indicator,* 0.1% solution, Dissolve 0.1 g of methyl red in 100 ml of 95% ethyl alcohol. Dispense from a dropper bottle.

270

750 ml	*Nitric acid,* 3 M. See p. 268. Dispense from several dropper bottles.
500 ml	*Phosphate control,* 0.4 mg P/5 ml. Dilute 12.9 ml of 0.1 M Na_2HPO_4 (see reagents for Expt. 7, **p. 274**) to 500 ml with water. (Can also use 0.1 M H_3PO_4 or NaH_2PO_4.)
150 ml	*Silver nitrate,* 0.5% solution. Dissolve 0.5 g $AgNO_3$ in 100 ml of 1 M HNO_3. Dispense from a dropper bottle.
300 ml	*Sodium hydroxide,* 10% solution. Dissolve 50 g NaOH pellets in water with *constant stirring* and then dilute to 500 ml. Store in polyethylene bottle.

Experiment 3.   Inorganic Elements of Biological Importance. Dry Ashing

500 g	*Dry skim milk* or *dry whole milk.*
500 g	*Oatmeal.*
3-6	*Balances.* Triple-beam, dial type, or top-loading.
200	*Dishes,* porcelain evaporating, 75 mm (locker).
200	*Filter paper,* medium grade, 11 cm.
20	*Tongs,* metal.
40	*Triangles,* nichrome wire, size C, 2½ inches side length, available on reagent shelf, for supporting porcelain evaporating dishes.
50 ml	*Acetic acid,* 1 M. See p. 268. Dispense from a dropper bottle.
400 ml	*Barium chloride,* 10% solution. Dissolve 50 g $BaCl_2$ in water and dilute to 500 ml.
200 ml	*Disodium phosphate,* 10% solution. Dissolve 50 g Na_2HPO_4 or 94 g $Na_2HPO_4 \cdot 7H_2O$ and make to 500 ml. Adjust to pH 8-8.5 with 3 M HCl or 3 M HNO_3.
400 ml	*Hydrochloric acid,* 3 M. See p. 268.
1 liter	*Nitric acid,* 3 M. See p. 268.
25 ml	*Potassium permanganate,* 0.5% solution. Dissolve 0.5 g $KMnO_4$ in 100 ml water. Dispense from a dropper bottle.
400 ml	*Potassium thiocyanate,* 10% solution. Dissolve 50 g KSCN in water and dilute to 500 ml.
NOTE:	See Expt. 2 p. 270-1, for preparation of the following reagents:
200 ml	*Ammonium chloride,* 10% solution.
1 liter	*Ammonium hydroxide,* 2 M.
1200 ml	*Ammonium molybdate solution.*
200 ml	*Ammonium oxalate,* saturated solution.
25 ml	*Methyl red indicator.*
50 ml	*Silver nitrate,* 0.5% solution.

Experiment 4.   Basic Principles of Colorimetry. The Phenanthroline Method for Iron

5-10	*Spectrophotometers.* Bausch and Lomb Spectronic 20 or similar type.
10-20	*Cuvettes.* Matched tubes, 12 mm (½″).
800 ml	*Hydroxylamine hydrochloride,* 10% solution. Dissolve 100 g $NH_2OH \cdot HCl$ in distilled water and make to 1 liter.
1.6 liters	*Orthophenanthroline,* 0.1% solution. Dissolve 1.0 g 1,10-phenanthroline monohydrate in distilled water with stirring and heating. *Do not heat over 80°C.* Store in a dark bottle; discard the solution if it becomes colored.
	(1 ml of 0.1% phenanthroline will complex no more than 100 μg Fe.)

800 ml

Sodium acetate, 1.5 M solution. Dissolve 204 g of $CH_3COONa \cdot 3H_2O$ in water and make to 1 liter.

1 liter

Stock iron solution, 1 ml = 200 μg Fe. Prepare from either ferrous ammonium sulfate or iron wire.

(a) Slowly add 20 ml conc H_2SO_4 into 50 ml of distilled water, and dissolve 1.404 g of analytical grade ferrous ammonium sulfate, $Fe(NH_4)_2(SO_4)_2 \cdot 6H_2O$, in this solution. Add 0.5% $KMnO_4$ dropwise until a faint pink color persists. Transfer the solution quantitatively to a 1 liter volumetric flask and dilute to the mark with iron-free water.

(b) If iron wire is used to prepare the stock solution, use electrolytic-grade, or "iron wire for standardizing." If necessary, remove any oxide coating with fine sandpaper. Weigh 0.200 g of clean wire and transfer to a 1 liter volumetric flask. (NOTE: Wear protective glasses when cutting wire to prevent eye injury from small chips.) Dissolve in 20 ml of 6 N H_2SO_4 and dilute to the mark with iron-free distilled water.

2.25 liters

Standard iron solution, 1 ml = 10 μg Fe. Pipet exactly 50.0 ml of "stock Fe solution" into a 1 liter volumetric flask and dilute with iron-free water to the mark.

NOTE: To save on reagents and to easily and accurately dispense the solutions, set up a pair of 25-ml burets for each reagent (except the stock Fe solution). Put a supply of each reagent in *clearly labelled* polyethylene "wash bottles" in order to readily fill the burets.

Analytical grade reagents with the lowest iron content should be used. Preferrably, glassware should be cleaned with 6 M HCl and thoroughly rinsed. Students should be made aware of the necessity to avoid iron contamination when determining "trace metals."

– –

Unknown Iron Solutions. Prepare a series of unknown Fe solutions for student unknowns in the range of 15 to 75 μg Fe per ml. Prepare as many different unknowns as desired by diluting the *stock* Fe solution (1 ml = 200 μg Fe). For example, to prepare 100 ml of unknown containing 15 μg Fe/ml (1500 μg Fe/100 ml) measure exactly 7.5 ml (1500 μg $\div$ 200 μg/ml = 7.5 ml) of the stock solution into a 100 ml volumetric flask and dilute to the mark. Each student should be given about 10 ml of a particular unknown in a test tube.

Experiment 5. Colorimetric Determination of Phosphorus and Iron. Wet Ashing

– –

Balances, analytical.

5-10

Spectrophotometers, Bausch and Lomb Spectronic 20 or similar type.

– –

Samples. Dry skim milk, oatmeal, soybean flour, whole wheat flour, liver or other substances with reasonable P and Fe contents. Consult references at end of Expt. 1. If sample has high water content, pre-dry the weighed sample (in 125-ml Erlenmeyer flask) at 105-110°C.

A. Preparation of Sample

100

Filter paper, Whatman No. 41, 11 cm.

20

Funnel, glass, short-stem. For digestion process (reagent shelf).

– –

Funnel, glass long-stem. For filtration after digestion (locker).

20

Funnel support, wooden.

20

Volumetric flask, 50 ml (locker or reagent shelf).

5 liters	*"Extraction Solution,"* acetic acid/sodium acetate. Dissolve 100 g sodium acetate, $CH_3COONa \cdot 3H_2O$, in distilled water, and then add 30 ml of glacial acetic acid and make to 1 liter.
750 ml	*Hydrogen peroxide,* 30% solution. *CAUTION:* Very corrosive to the skin. Handle carefully.
500 ml	*Sulfuric acid,* concentrated. A 25-ml dispensing buret on the 5-pint commercial bottle is recommended. See p. 268.

B. Phosphate Determination

1 liter	*Phosphate Standard,* 1 ml = 100 μg P. Dissolve 439.3 mg anhydrous KH_2PO_4 (previously dried @ 105° for 1 hr) in distilled water in a scrupulously clean 1-liter volumetric flask and make to 1 liter. (NOTE: Flask must be *thoroughly* rinsed if cleaned with a phosphate detergent.)
800 ml	*Acid molybdate reagent,* 2.5% ammonium molybdate in 5 N H_2SO_4. *Cautiously* add 136 ml of conc. H_2SO_4 to 360 ml of water with stirring. Cool the solution. In 500 ml of water, dissolve 25 g ammonium molybdate, $(NH_4)_6Mo_7O_{24} \cdot 4H_2O$, and add this solution to the sulfuric acid solution with stirring. Dispense from 10-ml buret on reagent bench.
800 ml	*Reducing agent: 1% Elon/3% sodium bisulfite.* Dissolve 15 g $NaHSO_3$ and 5 g of p-methylaminophenol (Elon) in 500 ml of water. Stable 10 days in brown bottle. Dispense from 10-ml buret on reagent shelf.

C. Iron Determination

Note:	See Expt. 4, p. 271-2 for the preparation of the following reagents. (The first 3 reagents should each be dispensed from 25-ml burets on reagent bench.)
800 ml	*Hydroxylamine hydrochloride,* 10% solution.
1600 ml	*Orthophenanthroline,* 0.1% solution.
800 ml	*Sodium acetate,* 1.5 M solution.
1 liter	*Stock iron solution,* 1 ml = 200 μg Fe.
2 liters	*Standard iron solution,* 1 ml = 10 μg Fe.

Experiment 6.   Active Acidity. Measurement of pH

2	*Hydrion pH test paper, wide range,* pH 0-13, with dispenser. Refills also needed.
1 each range	*Hydrion pH test paper, narrow range,* with dispensers. The following narrow ranges are recommended: 1.0-2.5, 3.0-5.5, 6.0-8.0, 8.0-9.5, 10.0-12.0, and 12.5-14.0. Refills are also needed.
4-5	*pH demonstration samples.* Dropper bottles containing 0.1 M HCl, 0.1 M NaOH, and 2-3 other liquids selected by instructor from list on p. 45.
5-10	*pH meters.*
1 liter	*Standard pH solutions,* pH 4.0 and 7.0 for standardizing pH meters. Available from various suppliers or can be prepared in accordance to published tables. (See references by Roger Bates, p. 45).
500 ml each	*Unknown pH samples.* Each student will be given about 20-25 ml of an unknown pH sample in a 18 x 150 mm test tube. A series of ten pH unknowns consisting of 0.05 M phosphate buffers can be prepared as described below:

(1) Prepare the following stock solutions:

Phosphoric acid, 0.5 M. Weigh 28.8 g of 85% H_3PO_4 in a 100-ml beaker, transfer to a 500-ml volumetric flask with water and dilute to mark.

Dihydrogen phosphate, 0.5 M. Dissolve 34 g KH_2PO_4 (or 34.5 g $NaH_2PO_4 \cdot H_2O$) in water and dilute to 500 ml.

Monohydrogen phosphate, 0.5 M. Dissolve 43.6 g K_2HPO_4 (or 35.5 g Na_2HPO_4) in water and dilute to 500 ml.

Sodium hydroxide, 1 M. Dissolve 40 g NaOH pellets in water with *constant stirring* and dilute to 1 liter.

(2) Preparation of individual unknown buffers (labeled A-J).
Unknowns A-C, pH 1.8 to 3.8.

Unknown A—Add 50 ml 0.5 M H_3PO_4 (graduate) into a graduated 600-ml beaker and add water to 500-ml mark. Mix and measure pH to closest 0.01 pH unit.

Unknown B—Add 50 ml 0.5 M H_3PO_4 to 600-ml beaker and dilute with water to approximately 450 ml. Place on a magnetic stirrer and lower pH electrodes into the solution. Add 1 M NaOH from a buret until the pH is about 2.5 ± 0.2. Add water to the 500-ml mark on beaker and read pH accurately to closest 0.01 pH unit.

Unknown C—Repeat as for unknown B but bring final pH to 3.5 ± 0.2.

Unknowns D-G, pH 4.1 to 8.5.

Unknown D is prepared as described for Unknown A, but using 0.5 M dihydrogen phosphate solution. Unknowns E-G are prepared with dihydrogen phosphate solution and 1 M NaOH as described for B above. Each successive solution should be approximately 1 pH unit higher than the previous one.

Unknowns H-J, pH 9 to 12.

Prepare with 0.5 M monohydrogen phosphate solution as described above, that is, H is similar to D, and I-J are similar to E-G. The unknowns can be stored in 500-ml screw-cap bottles and the unused portions can be kept for future use if a crystal of thymol or about 0.2 ml of octanoic (caprylic) acid is added to each bottle to prevent mold formation.

Experiment 7. Buffers. The Control of pH

5-10	*pH Meters.*
1 liter	*Acetic acid*, 0.1 M. Dilute 100 ml of 1.0 M acetic acid to 1 liter. See note below.
750 ml	*Ammonium chloride*, 0.1 M. Dissolve 5.35 g NH_4Cl in water and dilute to 1 liter.
600 ml	*Ammonium hydroxide (Ammonia)*, 0.1 M. Dilute 100 ml 1.0 M NH_4OH to 1 liter. See note below. Store in tightly-stoppered bottle!
250 ml	*Phosphoric acid*, 0.1 M. Pipet 100 ml of 0.5 M H_3PO_4 (see Expt. 6 above) into a 500-ml volumetric flask and dilute to mark. Alternate method: Weigh 5.8 g of 85% H_3PO_4 in a small beaker, transfer to a 500-ml volumetric flask with water and dilute to mark.
1.3 liters	*Sodium acetate*, 0.1 M. Dissolve 13.6 g of $CH_3COONa \cdot 3H_2O$ in water and dilute to 1 liter.
1.4 liters	*Sodium dihydrogen phosphate*, 0.1 M. Dissolve 13.8 g of $NaH_2PO_4 \cdot H_2O$ in water and dilute to 1 liter.
1 liter	*Sodium monohydrogen phosphate*, 0.1 M. Dissolve 14.2 g of anhydrous Na_2HPO_4 (or 26.8 g of $Na_2HPO_4 \cdot 7H_2O$) in water and dilute to 1 liter.
1.3 liters	*Tris.* 0.1 M. Dissolve 12.11 g of tris (hydroxymethyl)aminomethane in water and dilute to 1 liter. (Aldrich Chem. Co. or Sigma Chem. Co.)

1 liter	*Tris hydrochloride (Tris HCl)*, 0.1 M. Dissolve 15.76 g of tris(hydroxymethyl)-aminomethane hydrochloride in water and dilute to 1 liter. Alternate method: Dissolve 12.11 g of Tris in water. Pipet exactly 100 ml of 1.00 N HCl into this solution and dilute to 1 liter.

NOTE: If the students are to obtain the buffer pH assigned to them, all of the above 0.1 M solutions must be prepared with reasonable care. The preparation of *standardized* 1.0 M solutions of acetic acid and ammonium hydroxide is recommended (see page 269) and these solutions can be diluted to 0.1 M as described above.

2 liters	*Hydrochloric acid*, 0.1 M. See p. 269.
2 liters	*Sodium hydroxide*, 0.1 M. See p. 269.

Experiment 8. Total Acidity

— —	*Balances, analytical.*
20-100	*Burets*, 25-ml and 50-ml. Need 20 burets of each size if available from reagent bench or stockroom; need 100 each buret per 100 students if in individual lockers.
20	*Buret clamps.* Available on reagent bench.
100 each	*Pipets*, 5-ml and 10-ml. Usually in student locker.
— —	*Methyl red–methylene blue indicator (Tashiro's indicator).* Dissolve 0.25 g of methylene blue and 0.375 g of methyl red in 300 ml of 95% ethyl alcohol. Dispense from several dropper bottles.
— —	*Methyl orange indicator*, 0.1% solution. Dissolve 0.1 g of methyl orange in 100 ml of water. Dispense from a dropper bottle.
— —	*Phenolphthalein indicator*, 1% solution. Dissolve 2 g of phenolphthalein in 120 ml of 95% ethyl alcohol and dilute to 200 ml with water with stirring. Dispense from 3-4 dropper bottles.
20 liters	*Sodium hydroxide*, 0.100 to 0.125 N. See p. 269 for preparation and standardization. The *exact* normality should be given to the students.
20 liters	*Sulfuric acid*, 0.100 N. See p. 269.

NOTE: The volumes of standard NaOH and H_2SO_4 required for the experiment depend on size of the class and whether all parts of the experiment are performed. It is best to prepare a sufficiently large batch of each reagent so that the normalities are constant for the entire experiment. Large polyethylene jugs (1 to 5 gal) are available with spigots for conveniently filling reagent bottles. Students can be given 150-200 ml of NaOH in their 500-ml plastic bottle and the H_2SO_4 can be placed in a 1-liter bottle on the reagent shelf.

Samples of Common Substances (Part B)

2 liters	(1) *Household ammonia* (diluted 1 to 10). Mix 100 ml household ammonia with 900 ml distilled water. Make a large enough batch to serve all students to avoid change of NH_3 concentration during a class period. Store in rubber-stoppered bottle to prevent escape of NH_3.
— —	(2) *Vinegar*
— —	(3) *Fruit juices:* lemon or lime, grapefruit, white grape juice, or other juices.
— —	(4) *Solid substances:* cream of tartar, baking soda, antacid tablets, or other substances.
— —	*Unknown acids for equivalent weight (Part C).* The following solid acids are suggested: adipic, fumaric, glutaric, maleic, malic, malonic, oxalic, phthalic, succinic, tartaric, and potassium bitartrate.

Experiment 9. General Color Tests for Carbohydrates

100 ml	*Glucose*, 0.5% solution. ⎱ Dissolve 0.5 g of the sugar in
100 ml	*Sucrose*, 0.5% solution. ⎰ water and make to 100 ml.

100 ml   *Starch*, 0.5% solution. Slurry 0.5 g of soluble starch with a few ml of water and pour into 100 ml of boiling water. Boil 1 minute.

50 ml   *Molisch reagent*, 5% α-naphthol. Dissolve 5 g of α-naphthol in 100 ml 95% ethyl alcohol.

Note: The above three carbohydrates and Molisch reagent should be dispensed from dropper bottles.

900 ml   *Anthrone reagent*, 0.2% solution. Dissolve 0.2 g anthrone in 100 ml of 95% sulfuric acid. (Prepare 95% H_2SO_4 by cautiously adding 100 ml conc. H_2SO_4 to 5 ml distilled H_2O with cooling.) Reagent is stable about one week.

NOTE: The volumes of the above reagents to be prepared will depend on whether the instructor demonstrates the experiment or has the students perform it themselves.

— —   *Sulfuric acid*, concentrated. A 25-ml dispensing buret on the 5-pint commercial acid bottle is recommended. See page 268.

Experiment 10. Carbohydrate Tests Involving Oxidation of Sugars

200   *Dishes*, porcelain evaporating, 75-mm (locker).

4.5 liters   *Barfoed's reagent.* Dissolve 66.5 g of neutral, crystalline copper acetate in 800 ml of water, add 6 ml of glacial acetic acid and dilute to 1 liter.

4.5 liters   *Benedict's reagent.* Dissolve 173 g of sodium citrate dihydrate and 100 g of anhydrous sodium carbonate or 118 g $Na_2CO_3 \cdot H_2O$ in 800 ml of distilled water, filter if necessary, add slowly with stirring 17.3 g $CuSO_4 \cdot 5H_2O$ dissolved in 100 ml water, and dilute to 1 liter. This reagent is stable on long standing.

50 ml   *Hydrochloric acid,* 3 M. See page 268. Dispense from dropper bottle.

1 liter   *Nitric acid*, concentrated. A 50-ml dispensing buret on the 5-pint commercial bottle is recommended. See page 268.

1.2 liters   *Starch,* 1% solution. In a small beaker add sufficient water to 10 g soluble starch to make a thin slurry. Add the starch slurry to about 800 ml of boiling water and boil 1 minute. Cool and make to 1 liter.

1.2 liters   *Sucrose,* 1% solution. Dissolve 5 g sucrose in 500 ml water.

700 ml of each   *Sugars,* 1% solutions of glucose, fructose, galactose, lactose, maltose, and xylose. Prepare individual sugar solutions by dissolving 5 g of the sugar in 500 ml water.

1 liter   *Galactose,* 2% solution for mucic acid test. Dissolve 10 g galactose in water and make to 500 ml.

250 ml of each   *Unknown sugars,* 2% solutions. Make individual solutions of:

1. glucose	7. xylose
2. fructose	8. fructose + xylose
3. galactose	9. galactose + xylose
4. sucrose	10. galactose + fructose
5. lactose	11. sucrose + lactose
6. maltose	12. sucrose + maltose

Dissolve 5 g of the sugar in water and dilute to 250 ml. In mixtures 8-12, *each* sugar is present in 2% concentration. Each student will need 20-25 ml of the unknown in a 150-mm test tube.

NOTE: To prevent mold formation, all sugars (including students' unused unknowns) should be refrigerated when not in actual use. Solid sugars should be purchased in a minimum of 100 g quantities.

Experiment 11. Color Tests for Ketoses and Pentoses

150 ml	*Aniline acetate solution.* Shake 25 ml aniline with 25 ml of water and add glacial acetic acid (about 25 ml) until the solution clears. Dispense in dropper bottles. Keep tightly closed when not in use. Discard unused reagent.
2 liters	*Hydrochloric acid,* concentrated. A 100-ml dispensing buret on the 5-pint commercial acid bottle is recommended. See page 268.
3 liters	*Orcinol reagent.* Dissolve 0.5 g hydrated ferric chloride, $FeCl_3 \cdot 6H_2O$, and 0.5 g orcinol in 500 ml conc. HCl, and add this solution to 500 ml of water.
3 liters	*Seliwanoff's resorcinol reagent.* Dissolve 1 g of resorcinol in 1 liter of 4 M HCl (333 ml conc. HCl in 667 ml water).
1 pkg.	*Shredded wheat.*
50 ml	*Sorbose,* 1% solution. Dissolve 0.5 g sorbose in 50 ml of water. Place in dropper bottle.
500 ml of each	*Sugars,* 1% solutions of fructose, glucose, sucrose, xylose and starch. See Experiment 10, Reagents, for preparation.
25 ml of each	*Sugars,* in dropper bottles for Seliwanoff and orcinol tests. Use 1% sugar solutions listed above.

Experiment 12. Summary of Sugar Tests. No reagents required.

Experiment 13. Detection of Glucose in Urine

100	*Tubes, disposable culture.* 13 x 100 mm.
400	*CLINITEST tablets* (Ames). Available in bottles of 100 or 250 tablets from Scientific Products, a division of American Hospital Supply Corp., Evanston, Ill. Can be purchased from most drugstores.
8 btls.	*CLINISTIX* (Ames). Available as above.
3 rolls	*TES-TAPE* (Eli Lily). Available in drugstores. Either one of the above two test papers is required.
– –	*Distilled water* in 2 dropper bottles.
100 ml	*Glucose,* 1% solution. 1 g glucose per 100 ml water. Dispense from 2 dropper bottles.
– –	*Urine.* Fresh specimen daily in 2 dropper bottles.
50 ml	*Vitamin C (ascorbic acid),* 0.5% solution. 0.5 g ascorbic acid made to 100 ml with water. Dispense from 2 dropper bottles. Keep refrigerated when not in use. Make fresh weekly.

Experiment 14. Determination of Glucose in Blood Plasma

5-10	*Spectrophotometers.* Bausch and Lomb Spectronic 20 or similar type.
10-20	*Cuvettes.* Matched tubes, 12 mm (½″).

– –	*Blood Plasma.* Drawn blood, or *Lab-trol* and *Patho-trol.*
– –	(1) *Anticoagulant and Preservative for Drawn Blood.* $K_2C_2O_4$ and NaF are available in tablet form to use when collecting blood with a syringe, or these reagents can be obtained in proper quantities in "Vacutainer Specimen Tubes." (Scientific Products, catalog 1978, Division of American Hospital Supply Corp., Evanston, Ill., pp. 59-62)
2 x 7.5 ml vials *Lab-trol*	(2) *Lab-trol* and *Patho-trol.* These two commercial products (Dade) contain solutions of plasma proteins, glucose, and many organic and inorganic substances of known amounts (Scientific Products, catalog 1978, pp. 104-5).
6 x 7.5 ml vials *Patho-trol*	*Lab-trol* and *Patho-trol* can be used as "glucose unknowns" by diluting them with water in various proportions. Alternatively, in case of budgetary necessity, a series of dilute glucose solutions can be prepared as "unknowns."
100 ml	*Glucose, 1% stock solution.* Weigh exactly 1.000 g of pure glucose. Transfer to a 100-ml volumetric flask, dissolve in 0.1% benzoic acid solution and dilute to the mark with the same solvent.

Somogyi-Nelson Method for Glucose (parts A and B)

100	*Filter paper,* Whatman No. 40, 11 cm.
150	*Sugar tubes, Folin-Wu,* 25 ml.
200 ml	*Zinc Sulfate,* 5% solution. Dissolve 50 g $ZnSO_4 \cdot 7H_2O$ in water and dilute to 1 liter. Dispense from a 25-ml buret.
200 ml	*Barium hydroxide,* 0.3 N solution. Dissolve 45 g $Ba(OH)_2 \cdot 8H_2O$ in water and dilute to 1 liter. Filter the solution, if cloudy, through a sintered-glass funnel under vacuum using a water aspirator. Store in an aspirator bottle stoppered with a soda lime tube to prevent absorption of carbon dioxide. Pipet 10.0 ml of $ZnSO_4$ solution into a 250-ml Erlenmeyer flask, add 50 ml of water and titrate slowly with $Ba(OH)_2$ in the presence of 4 drops of 1% phenolphthalein solution to faint pink endpoint. Adjust with water (preferably the $ZnSO_4$ solution) so that 10.0 ml of the $ZnSO_4$ solution reacts with 10.0 ml of the $Ba(OH)_2$ solution. Repeat the titration (and adjustment) until the two solutions balance stoichiometrically. Dispense from a 25-ml buret with a soda lime tube attached to the top of the buret. **NOTE:** Balanced solutions of 5% $ZnSO_4$ and 0.3 N $Ba(OH)_2$ are available from Sigma Chemical Co., St. Louis, Mo.
115 ml daily	*Copper reagent.* Prepare a *fresh* mixture of 100 ml of reagent A and 15 ml of reagent B. Discard left-over reagent.
500 ml	*Reagent A.* In a 1-liter volumetric flask containing about 600 ml distilled water dissolve, in turn, 25 g anhydrous Na_2CO_3, 25 g $NaKC_4H_4O_6 \cdot 4H_2O$ (Rochelle salt), 20 g $NaHCO_3$, and 200 g anhydrous Na_2SO_4. Dilute to mark. Filter if necessary and store at a temperature above 20°C.
75 ml	*Reagent B.* Dissolve 4 g $CuSO_4 \cdot 5H_2O$ with distilled water in a 100-ml volumetric flask and dilute to the mark.
700 ml	*Arsenomolybdate reagent.* Dissolve 50 g of ammonium molybdate in 800 ml of water. Add 42 ml of concentrated sulfuric acid and mix. Add 6 g $Na_2HAsO_4 \cdot 7H_2O$ (or 2.8 g H_3AsO_4) dissolved in 50 ml of water. Mix and place in an incubator (or water bath) at 37°C for one to two days. Store in a brown glass-stoppered bottle. **NOTE:** The copper reagent mixture (A + B) and the arsenomolybdate reagent should be dispensed from 10-ml burets on the reagent bench. A supply of each

reagent should be available in *clearly-labeled* polyethylene "wash bottles" for filling the burets.

— —   *Glucose, dilute solution for standards,* (0.20 mg/ml). Pipet 2.00 ml of the 1% stock solution (see p. 278) into a 100-ml volumetric flask. Dilute to the mark with 0.1% benzoic acid solution as a preservative. Store in refrigerator. Discard after one week.

Glucose-Oxidase Method for Glucose (part C)

— —   *Glucose, dilute solution for standards,* (0.10 mg/ml). Pipet exactly 1.00 ml of the 1% stock solution (see p. 278) into a 100-ml volumetric flask and dilute to the mark with a 0.1% benzoic acid solution as a preservative. Store in the refrigerator. Discard after one week.

2.5 liters (25 caps.)   *Glucose-Oxidase/Peroxidase enzymes.* Available from Sigma Chemical Co., P.O. Box 14508, St. Louis, Mo., 63178. Prepare as needed. Dissolve the contents of five capsules in 500 ml distilled water in an amber bottle. Mix gently. Store in refrigerator. Each capsule contains 500 International Units of glucose oxidase and 100 purpurogallin units of peroxidase.

40 ml   *o-Dianisidine hydrochloride solution.* Dissolve 50 mg o-dianisidine hydrochloride in 20 ml distilled water. Stable in refrigerator for 3 months. Available from Sigma Chemical Co. in preweighed vials.

2.5 liters   *Combined Enzyme-Color reagent.* Add 8.0 ml of o-dianisidine solution to 500 ml of the glucose oxidase/peroxidase solution. Solution is stable for one month in the refrigerator. Dispense from 50-ml buret on reagent bench. Use a *clearly-labeled* polyethylene "wash bottle" for filling buret. Keep refrigerated when not in use.

Experiment 15.   Quantitative Determination of Sugars in Foods

40   *Aluminum foil,* one inch squares. For honey and syrups.

— —   *Balances,* analytical.

100   *Filter paper,* 11 cm, medium grade.

25   *Flasks,* volumetric, 50 ml. Check out, and return to stockroom or reagent shelf at end of lab period.

25   *Flasks,* volumetric, 250 ml. Check out, and return to stockroom or reagent shelf at the end of lab period.

— —   *Foods.* List to be supplied by instructor.

50   *Test tubes,* 25 x 200 mm. Check out, and return to stockroom or reagent shelf at end of lab period.

250 ml   *Cupric sulfate,* 5% solution. Dissolve 5 g of anhydrous $CuSO_4$ (or 7.8 g $CuSO_4 \cdot 5H_2O$) in water to make 100 ml.

1 liter   *Copper reagent.* Dissolve the following chemicals in the order given in about 800 ml of water: 7.5 g $CuSO_4 \cdot 5H_2O$; 25 g $NaKC_4H_4O_6 \cdot 4H_2O$ (Rochelle salt); 25 g Na_2CO_3; 20 g $NaHCO_3$; 5.0 g KI. Accurately weigh 0.7850 g of KIO_3, dissolve in 20-30 ml water, rinse quantitatively into the main solution, and dilute with water to 1 liter. Mix thoroughly. Dispense from 50 ml burets on reagent shelf; fill burets from clearly labelled polyethylene "wash bottle."

50 ml   *Phenolphthalein indicator,* 1% solution, see Expt. 8, p. 275.

1 liter   *Sodium hydroxide,* 1 M solution. See p. 268.

500 ml   *Sodium hydroxide,* 0.1 M solution. See p. 268.

1 liter	*Sodium thiosulfate*, 0.1 N solution. Dissolve 25 g of $Na_2S_2O_3 \cdot 5H_2O$ in water and dilute to 1 liter. Standardize against 0.100 N dichromate solution (4.903 g of pure, dry $K_2Cr_2O_7$ per liter) as follows: pipet 25 ml of 0.100 N dichromate solution into a 250-ml Erlenmeyer flask, add 25 ml of 1 N H_2SO_4 and 3 g KI. Mix, let stand 1 minute, and titrate the liberated iodine with the thiosulfate solution. Add starch indicator just before the last of the iodine color is discharged, and titrate to the disappearance of the blue color and only the green color of the chromic ion remains. Repeat the above to obtain an average normality value of the thiosulfate solution. If desired, the thiosulfate solution can be adjusted to exactly 0.100 N by adding a calculated amount of water to the solution whose normality usually comes out greater than 0.1 N, and then re-standardize to check if the adjusted solution is now 0.100 N.
3 liters	*Sodium thiosulfate*, 0.005 N for student use. Pipet 50.00 ml of 0.100 N sodium thiosulfate (or the exact calculated volume if the thiosulfate solution is *not* 0.100 N) into a 1-liter volumetric flask and dilute to the mark with water. **NOTE:** The 0.005 N thiosulfate solution does not keep for more than a few hours and, therefore, must be *freshly* prepared immediately before use and discarded after the lab period.
75 ml	*Starch indicator solution.* Slurry 1 g of soluble starch in 5 ml of water and pour this suspension into 95 ml of boiling water. Stir and boil for about 3 minutes, then cool and add 0.1 g $ZnCl_2$ as a preservative. Dispense from dropper bottles.
3 liters	*Sulfuric acid*, 0.5 M solution, see p. 268. *Sugars for calibration curves:* *Lactose*
100 ml	(1) *Stock lactose solution.* Transfer 1.500 g of pure anhydrous β-lactose (or 1.580 g of α-lactose monohydrate) into a 100-ml volumetric flask, dissolve with water and dilute to the mark. Refrigerate when not in use.
100 ml	(2) *Dilute lactose solution.* Pipet 5.00 ml of stock lactose solution into a 100-ml volumetric flask and dilute to mark. 1 ml = 0.75 mg lactose. Solution must be *freshly* prepared when needed. *Glucose*
100 ml	(1) *Stock glucose solution.* Transfer 0.9 g of pure, anhydrous glucose to a 100-ml volumetric flask. Dissolve in water, add 1 ml of 0.5 M H_2SO_4, mix and dilute to the mark. Refrigerate when not in use.
100 ml	(2) *Dilute glucose solution.* Pipet 5.00 ml of the stock solution into a 100-ml volumetric flask and dilute to the mark. 1 ml = 0.45 mg glucose. Solution must be *freshly* prepared when needed.

Experiment 16.   Esters

400	*Filter paper*, 11 cm, medium grade.
Bottle	*Glass wool.*
200 ml	*Acetic acid*, glacial.
400 ml	*Acetic anhydride.*
35	*Aspirin tablets.* Crushed and stored in tightly closed bottle.
200 ml	*Ethyl alcohol*, absolute.
200 ml	*Ethyl butyrate.*

50 ml	*Ferric chloride,* 0.1 M solution. Dissolve 1.35 g $FeCl_3 \cdot 6H_2O$ in 50 ml water. Dispense from a dropper bottle.
300 ml	*Hydrochloric acid,* 3 M. See p. 268.
200 ml	*Methyl alcohol,* anhydrous.
20 ml	*Phenolphthalein indicator,* 1% solution. See Expt. 8, p. 275.
300 ml	*Potassium hydroxide,* 3 M solution. Dissolve 19.8 g KOH pellets (85% assay) in 75 ml of 50% ethyl alcohol (alcohol:H_2O, 1:1 by vol.) with constant stirring and cooling. Make to 100 ml with 50% ethyl alcohol and store in a small polyethylene bottle. Prepare fresh as needed; if solution darkens, discard it.
300 g	*Salicylic acid.*
— —	*Sulfuric acid,* concentrated. A 25-ml dispensing buret on the 5-pint commercial acid bottle is recommended. See p. 268.

Experiment 17. Saponification of Fats

	Fats, solid. Beef tallow, butter, lard, oleomargarine, etc.
	Fatty oils. Corn, cottonseed, olive, peanut, safflower, soybean, etc.
100	*Filter paper,* 11 cm, medium grade.
1	*Hydrion test paper* in dispenser, pH 8-10.
50 ml	*Calcium chloride,* 0.2 M solution. Dissolve 3 g $CaCl_2 \cdot 2H_2O$ in 100 ml water. Dispense from a dropper bottle.
500 ml	*Chromotropic acid solution.* Dissolve 1 g of chromotropic acid (4,5-dihydroxy-2, 7-naphthalenedisulfonic acid disodium salt) in 100 ml water. To this aqueous solution slowly add with stirring 450 ml of conc. H_2SO_4. Stable for two weeks.
	Label: HANDLE WITH CARE—H_2SO_4
1500 ml	*Detergent, synthetic,* 5-10% solution. Select a commercial detergent which will give a clear solution. Dissolve 50 g of a solid detergent (e.g., laboratory Alconox), or 100 g of a liquid detergent in distilled water to make 1 liter of solution.
10 liters	*Ethyl alcohol,* 95%.
50 ml	*Lead acetate,* 0.2 M solution. Dissolve 7.6 g $Pb(C_2H_3O_2)_2 \cdot 3H_2O$ in 100 ml water. Dispense from a dropper bottle.
50 ml	*Magnesium chloride,* 0.2 M solution. Dissolve 4 g $MgCl_2 \cdot 6H_2O$ in 100 ml of water. Dispense from a dropper bottle.
20 ml	*Phenolphthalein indicator,* 1% solution. See Expt. 8, p. 275.
— —	*Soap,* Ivory, bar or flakes.
50 ml	*Sodium bisulfite,* 10% solution. Dissolve 10 g of $NaHSO_3$ or $Na_2S_2O_5$ in water to make 100 ml. Dispense from a dropper bottle.
3 liters	*Sodium chloride,* saturated solution. Gradually add about 360 g NaCl to 1 liter of water with constant mixing (magnetic stirrer) until no more salt dissolves. If undissolved salt remains, add small increments (10 ml) of water until clear.
500 g	*Sodium hydroxide,* pellets.
500 ml	*Sodium hydroxide,* 1 M solution. See p. 268.
3 liters	*Sodium hydroxide,* standard 0.1 N solution. See p. 269.
50 ml	*Sodium periodate,* 0.1 M solution. Dissolve 2.1 g $NaIO_4$ in 100 ml water. Dispense from a dropper bottle.
2 liters	*Sulfuric acid,* 3 M solution. Cautiously and slowly add 168 ml concentrated H_2SO_4 to 750 ml water. Cool and dilute to 1 liter.

Experiment 18. Nonsaponifiable Materials. Sterols and Tocopherols (Vitamin E)

150 ml	*Acetic anhydride.* Dispense from dropper bottle.
700 ml	*n-Butyl alcohol.*
200 ml	*Cholesterol,* 0.05% solution. Dissolve 0.10 g pure cholesterol in 200 ml of chloroform. Discard unused solution.
200 ml	*Ergosterol,* 0.05% solution. Dissolve 0.10 g pure ergosterol in 200 ml chloroform. Discard unused solution.
100 ml	*Nitric acid,* concentrated.
— —	*Sulfuric acid,* concentrated.
200 ml	*Tocopherol,* 0.05% solution. Dissolve 0.10 g α-tocopherol or the mixed isomers of tocopherol in 200 ml of chloroform. Discard unused solution.

Experiment 19. Determination of Cholesterol in Blood Serum

5-10	*Spectrophotometers,* Bausch and Lomb Spectronic 20 or similar type.
— —	*Blood serum or plasma*
	(1) *Drawn Blood* from students or other source.
	(2) *Commercial cholesterol controls.* Various products are available such as Choles-trol (Scientific Products, McGraw Hill Park, Ill., 1978 catalog, p. 106) and Lipid Control-N and Lipid Control-E (Sigma Chemical Co., St. Louis, Mo.). These assayed controls, as such or diluted to various levels, can be used as *Student Unknowns.*
1	*Water bath,* commercial type. If not available, students can use beaker of water kept at 37°.

Liebermann-Burchard Method (part B)

500 ml	*Cholesterol standard solution,* 2 mg cholesterol/ml. Make up in an appropriate-size volumetric flask at a concentration of exactly 0.200 g of pure cholesterol per 100 ml glacial acetic acid.
3 liters	*Color reagent,* mixture of glacial acetic acid, acetic anhydride, a d H_2SO_4 in a volume ratio of 45:45:10. Slowly add 450 ml of cooled acetic anhydride to 450 ml glacial acetic acid with constant stirring in a 2-liter beaker set in an ice-water bath. To this cooled mixture *slowly* add 100 ml conc. H_2SO_4 with stirring. Dissolve 10 g anhydrous sodium sulfate (1% w/v) in the mixture as a stabilizer. Reagent is stable for about two weeks at room temperature. Dispense from a 50-ml buret on reagent bench.
	NOTE: If color intensities are measured with a spectrophotometer, the following items will be needed:
— —	*Test tubes,* 18 x 150 mm (student locker). Six tubes to be matched by student.
— —	*Test tube holder (adapter)* ¾″. One adapter per spectrophotometer to hold 18 x 150 mm tubes.
— —	*Shields,* to cover test tube in holder. Use commercial shield, paper cup, or tin can.

Enzymatic Method (part C)

20-30	*Cuvettes,* 12 mm (½″), for spectrophotometer.
100	*Test tubes (culture tubes),* 13 x 100 mm, for unknown samples.
— —	*Enzyme reagent,* mixture of enzymes and color reagents. Available as Kit No. 350 from Sigma Chemical Co., P.O. Box 14508, St. Louis, Mo., 63178. The kit contains: cholesterol esterase, cholesterol oxidase, peroxidase, phenol, 4-amino-

antipyrine, and phosphate buffer. (Enzyme mixture may be available from other suppliers.)

—— *Cholesterol standard*, 200 mg/100 ml. A vial containing 2 ml of an aqueous saline solution of cholesterol is included in each Sigma Kit No. 350. Cholesterol standard can also be made by dissolving 200 mg of high purity cholesterol in 100 ml of isopropyl alcohol which is also available from Sigma Chem. Co. (See original article by Allain in references, p. 132.)

Experiment 20.   Unsaturation of Fats. Iodine Number

—— *Flasks, Erlenmeyer*, 250-ml. In student lockers.

50 *Stoppers, rubber*, No. 6. Return after use.

100 *Pipet, long tip, disposable.* (Scientific Products, cat. no. P5205).

25 *Bulb, rubber*, 2 ml. (Scientific Products, cat. no. R5002-2) Return after weighing samples.

1-2 *Pipet, automatic*, 25-ml, with Teflon stopcock (Kimble No. 37077-F). For dispensing iodine solution. Reservoir capacity is 2000 ml. Mount firmly on large ring stand using a sturdy, split iron ring.

2 liters *Chloroform.*

6 liters *Hanus iodine solution.* To obtain equivalent amounts of iodine and bromine in this reagent, the amounts of each halogen should be measured as closely as practical. Weigh 26.4 g of iodine to within ± 0.02 g on a top-loading or triple-beam balance using a small, tared glass beaker. Transfer the iodine to a 2-liter volumetric flask, add about 1800 ml of glacial acetic acid, and dissolve the iodine by frequent swirling, warming on a hot plate if necessary. (*Do not overheat*; warm to the touch.) Allow the iodine solution to cool to room temperature.

Carefully add 5.35 ml of bromine to the iodine solution.
(**CAUTION:** WEAR RUBBER GLOVES and WORK UNDER A GOOD VENTILATING HOOD. A recommended procedure is to *partially* fill a clean 10-ml or 25-ml buret (Teflon stopcock) with the bromine using a small glass funnel, and then carefully dispense the bromine from the buret into the 2-liter volumetric flask. Liquid bromine spilled on the skin can cause serious burns. Do not inhale fumes.)

Fill the volumetric flask to the 2-liter mark with glacial acetic acid. Stopper tightly and mix the solution.

2.5 liters *Potassium iodide*, 15% solution. Dissolve 150 g KI in about 800 ml of water and dilute to 1 liter.

6-8 liters *Sodium thiosulfate*, 0.1 N solution. See Expt. 15, page 280 for the preparation and standardization of 0.1 N $Na_2S_2O_3$.

500 ml *Starch indicator.* See Expt. 15, page 280.

—— *Student Unknowns* of fatty oils. Supply each student with 1-2 ml of a fatty oil in a small test tube. If a large number of samples must be issued, it is recommended to set up a series of dropper bottles of fatty oils from which the instructor can rapidly dispense 1-2 ml of student unknowns. The exact iodine number of each fatty oil can be determined by the instructor or from the average values obtained by a number of students.

The following list of fatty oils, with their iodine number range, are suggested: olive, 75-95; peanut, 85-100; cottonseed, 100-117; corn, 115-130; sunflower, 122-136; soybean, 125-140; safflower, 130-150; linseed, 170-205.

Experiment 21. Thin-Layer Chromatography of Lipids

> **NOTE:** If commercial silica gel chromatography sheets are only used, the first five items listed below will not be needed.

200   *Glass plates,* 40 x 90 mm. (Camag, Inc., 16229 Ryerson Road, New Berlin, Wis. 53151) If these plates are not available, microscope slides (25 x 75 mm) can be substituted, together with a 150-ml beaker containing 5 ml of the solvent system.

1 roll   *Adhesive tape.* Available at drug store, must be cloth type in order to be of suitable thickness and to tear into strips.

300 g   *Silica gel G,* (SilicAR 7G sorbent containing 14% plaster of Paris, Mallinckrodt Chemicals, available from most chemical suppliers) or equivalent grade of material.

600 ml   *Starch,* 3% solution. To 30 g of soluble starch add water to make a thin slurry and slowly add this to 600-700 ml of boiling water. Boil for 1 minute, cool and dilute to 1 liter. Add a few ml of toluene as a preservative.

100   *Scoopula,* for spreading silica gel coating. (Fisher Scientific Co.) Should be part of locker equipment.

1 roll   *Aluminum foil.*

15   *Chromatography sheets,* pre-coated silica gel, 100-micron thick coating, 20 cm x 20 cm, with polyacrylic acid binder (Eastman Chromatogram Sheet No. 13181, with fluorescent indicator). Equivalent brands must *not* be affected by iodine vapors. Cut each 20 x 20 cm sheet into eight 5.0 x 9.5 cm (2″ x 3¾″) sheets. Activate in oven at 100° for 15-30 minutes. Store in a desiccator.

100   *Filter paper,* 9 cm, medium grade.

200   *Micropipets,* one microliter (1 μl), "Microcap" pipet. (Drummond Science Co., 500 Parkway South, Broomall, Pa. 19008)

100 ml each   *Lipid samples,* 5 mg/ml in chloroform. Prepare *separate* lipid samples of lecithin, cholesterol, and vegetable oil (cottonseed or corn oil) by dissolving 0.5 g of the individual lipid in 100 ml of chloroform. Prepare a mixture of the three lipids by dissolving 0.25 mg of each in a single 50 ml chloroform solution. Depending on size of class, set up several labeled screw-cap vials of each of the four solutions for student use during spotting of chromatograms.

— —   *Lipid unknowns.* The above lipid samples can also be used as student unknowns. In addition, unknown mixtures of lecithin and cholesterol, lecithin and vegetable oil, and cholesterol and vegetable oil should be prepared at the 5 mg/ml concentration of each lipid.

1 liter   *Solvent system.* Petroleum ether, ethyl ether, and glacial acetic acid in volume ratio of 70:30:1. Mix 350 ml petroleum ether, 150 ml ethyl ether, and 5 ml glacial acetic acid. The petroleum ether used here is mainly hexane. (Skellysolve B, B.P. 60-70° C).

— —   *Iodine crystals.* Sprinkle a number of iodine crystals on the bottom of several 250-ml beakers and cover with watch glasses. Do *not* cover with aluminum foil. Place beakers in fume hood.

Experiment 22. Determination of Fat by Soxhlet Extraction

> **NOTE:** If this experiment is to be demonstrated, only a single assembly of the Soxhlet apparatus and an adjustable hot plate is required.

— —   *Balances,* analytical.

9 liters	*Chloroform,* reagent grade. The volume per 100 students is based on the extraction of *one* sample per student. Some of the chloroform can be recovered as "Used Chloroform."
20	*Desiccator.* See Expt. 1, p. 270.
6-12	*Extraction apparatus, Soxhlet,* small. The complete assembly consists of a condenser, extraction chamber, and flask, and the catalog numbers are: Pyrex No. 3840, Kimble No. 24005, and Kontes No. K-585000. The individual components of the apparatus are described as follows:

> Extractor condenser, Allihn type, 34/45 joint.
> Extraction chamber, 30 mm i.d. tube, 34/45 top joint, 24/40 bottom joint.
> Flask, boiling, flat-bottom, short neck 24/40 joint, 125-ml capacity.

NOTE: The extraction chamber with a stopcock drain is no longer listed as a stock item in the above catalogs. It probably can be special ordered. SGA Scientific Inc., 737 Broad St., Bloomfield, New Jersey 07003, lists it in the Inter-Joint section of their 1980 catalog, p. 1196, No. JE-8250.

box of 25	*Extraction thimbles,* paper, 25 x 80 mm, Whatman. Thimbles can be used repeatedly.
bottle	*Glass wool.*
1-2	*Hot plate,* six-place heaters, rheostat controlled. (Sargent-Welch S-32045, made by Precision Scientific Co.)
— —	*Unknown samples.* Ralston cereal (low fat) and wheat germ (high fat), individually and in blends of 25-75, 50-50, 75-25, etc. make excellent unknowns for this experiment.

Experiment 23. General Characteristics of Proteins. Comparison with Lipids and Carbohydrates

50 g	*Carbohydrate.* Sucrose, lactose, *or* glucose.
50 ml	*Fat.* Cottonseed *or* corn oil in a dropper bottle.
— —	*Proteins.* Suggested list: powdered casein, egg albumin, gelatin, soy protein, wheat gluten, hair or feathers.
2 vials	*Hydrion paper strips,* wide pH range.
500 ml	*Hydrochloric acid.* 3 M. See p. 268.
2 vials	*Lead acetate paper.*
50 g	*Soda lime.*

Demonstration (for parts B and C)

3	*Crucibles,* porcelain, 25 mm.
25 ml	*Chloroform.*
25 ml	*Hydrochloric acid,* 0.1 M. See p. 268.
25 ml	*Sodium hydroxide,* 0.1 M. See p. 268.

Experiment 24. Color Tests for Proteins and Amino Acids

A. General Protein Color Tests

600 ml	*Alanine,* 1% solution. Dissolve 5 g alanine in 500 ml of water. Store in refrigerator.
25 ml	*Ammonia, dilute,* 1 M solution. See page 268. Dispense from dropper bottle.
50 ml	*Copper sulfate,* 0.1% solution. See Expt. 2, p. 270. Dispense from dropper bottle.

600 ml   *Egg albumin (egg white),* 1% solution. Prepare an approximately 1% solution by mixing 1 part egg white with 10 parts of water and filtering through several folds of cheese cloth in a 5″ funnel. (Egg white contains 10-11% protein.) Store in refrigerator.

NOTE: Fresh egg white gives very much better test results than those obtained using powdered egg albumin.

1 liter   *Ninhydrin,* 0.1% solution. Prepare a fresh solution by dissolving 0.5 g ninhydrin in 500 ml water. Do not use old solutions from previous semester.

800 ml   *Nitric acid,* concentrated. Dispense from special acid buret. See page 268.

5-10 g   *Protein,* solid. Use casein or other solid protein.

500 ml   *Sodium hydroxide,* 10% solution. See Expt. 2, page 271.

600 ml   *Sucrose,* 1% solution. Dissolve 5 g sucrose in 500 ml water. Store in refrigerator.

300 ml   *Tryptophan,* 0.02% solution. Add 0.1 g tryptophan to 500 ml water and mix until completely dissolved. Store in refrigerator.

B. Color Tests for Specific Amino Acids

5-10 g   *Casein,* small bottle.

100 ml   *Cystine,* 0.1% in 1 M H_2SO_4 solution. Dissolve 0.2 g cystine in 200 ml 1 M H_2SO_4 (see page 268).

900 ml   *Egg albumin (egg white),* 1% solution. See part A above.

600 ml   *Gelatin,* 1% solution. Soak 10 g gelatin in a small amount of cold water, then add about 800 ml of water and warm, with stirring, until dissolved. Dilute to 1 liter.

500 ml   *Gelatin,* 0.1 % solution. Dilute 50 ml of 1% gelatin solution to 500 ml.

900 ml   *Hopkins-Cole reagent.* Place 10 g of powdered magnesium in a large Erlenmeyer flask and shake up with enough water to cover the magnesium liberally. Gradually add 250 ml of a cold, saturated solution of oxalic acid. The reaction proceeds very rapidly and with the liberation of much heat, so that the flask should be cooled under running water during the addition of the oxalic acid. The contents of the flask are shaken after the addition of the last portion of the acid and then poured upon a filter, to remove the insoluble magnesium oxalate. A little water is poured through the filter, the filtrate acidified with acetic acid to prevent the partial precipitation of the magnesium on long standing, and made up to 1 liter with water. This solution contains only the magnesium salt of glyoxylic acid.

25-50 g   *Lead acetate,* crystals in small bottle.

75 ml   *Millon's reagent.* Dissolve 1 part by weight of mercury with 2 parts by weight of conc. HNO_3 (sp. g. 1.42) and dilute the resulting solution with 2 volumes of water. **CAUTION:** Carry out the reaction in a fume hood and avoid inhaling the toxic brown fumes of NO_2. Use a large beaker because the reaction is somewhat violent. Add the nitric acid a little at a time.

    Example: Using an eyedropper, weigh about 10 g of mercury in a tared 250-ml beaker. For each 10 g Hg, slowly add 14 ml of conc. HNO_3 in small portions and, after the reaction is complete, add 30 ml of water.

Dispense from a dropper bottle. Freshly-prepared Millon's reagent works best.

100 ml   *α-Naphthol,* 0.02%. Dissolve 0.05 g of α-naphthol in a few ml of 95% ethyl alcohol and add 250 ml of water with constant stirring. Alternatively, if Molisch reagent (Expt. 9, page 276) is available, add 10-12 drops of this reagent to 100 ml of water.

300 ml   *Salicylic acid,* 0.02% solution. Dissolve 0.1 g of salicylic acid in 500 ml of water.

700 ml	*Sodium carbonate*, saturated solution. Add $Na_2CO_3 \cdot H_2O$ to 500 ml water with constant stirring until solution is saturated. This will take about 300 grams of the $Na_2CO_3 \cdot H_2O$. Do not use a glass-stoppered bottle.
100 ml	*Sodium hydroxide*, 10% solution. See part A above.
500 ml	*Sodium hydroxide*, 5% solution. Dissolve 25 g NaOH pellets in 400 ml of water with constant stirring. Dilute to 500 ml.
25 ml	*Sodium hypobromite (NaOBr)*. Carefully add 2 g (0.7 ml) of bromine (CAUTION!) to 100 ml of 5% NaOH. Keep in dark-colored bottle. Dispense from a dropper bottle. Stable for two weeks.
300 ml	*Sodium sulfite*, 20% solution. Dissolve 100 g of Na_2SO_3 in about 350 ml of water and dilute to 500 ml.
1.5 liters	*Sulfuric acid*, concentrated. Dispense from special acid buret. See page 268.
300 ml	*Tyrosine*, 0.02% solution. Since tyrosine is insoluble in water, add 1 N HCl dropwise to 0.1 g of tyrosine until it is completely dissolved. Then dilute to 500 ml with water.
100 ml	*Uric acid reagent (Folin's cystine reagent)*. Mix 20 g of sodium tungstate dihydrate with 8.5 ml of 85% H_3PO_4 and 30 ml of water. Reflux gently for 2 hours, add bromine water drop by drop until the solution is decolorized, remove the excess bromine by boiling, cool, and dilute to 100 ml.

Experiment 25. Separation and Identification of Amino Acids by Paper Chromatography

1 roll	*Aluminum foil.*
200	*Chromatographic paper*, Whatman No. 1, 12.5 x 23 cm (5″ x 9″). Whatman No. 1 paper comes in large 46 x 57 cm sheets with the grain of the paper (indicated by arrows) running in the longer dimension. Cut the paper into 12.5 x 23 cm sheets *so that the grain of the paper* runs in the direction of the 12.5 cm (5″) length. To do this, proceed as follows: Cut the paper thus—Fold 2-3 sheets in half the long way, crease and cut with a sharp knife. Cut these long pieces into four 12.5 cm (5″) widths. Eight small sheets will be obtained from 1 large sheet, and the grain of these cut sheets will be in the 12.5 cm direction. Avoid handling the paper excessively because the ninhydrin can show fingerprints.
100	*Filter paper*, 9 cm, medium grade.
100	*Micropipets*, one-microliter (1 μl), "Microcap" pipet. See Expt. 21, p. 284.
20	*Spot plate*. Part of locker equipment or available on reagent shelf or from stockroom.
1-2	*Stapler.*
1 liter	*Ninhydrin*, 0.2% solution in acetone. Dissolve 2 g ninhydrin in 1 liter of acetone.
2 liters	*Solvent A. 1-Butanol, glacial acetic acid, and water.* Volume ratio of 90:10:25. Do not use old solutions since butyl acetate may form on long standing.
2 liters	*Solvent B. 2-Butanol, 3.3% NH$_4$OH.* Volume ratio of 75:30. Stable if kept tightly closed to prevent loss of NH$_3$. The 3.3% NH$_4$OH is prepared by diluting 65 ml of concentrated NH$_4$OH to 500 ml with water. These solvent systems are prepared in accordance with the National Academy of Science reference listed on p. 177.
50 ml each	*Amino acid solutions*, 0.02 M solutions. Dissolve 0.001 mole of the amino acid in 1 ml of 1 M HCl and then dilute to 50 ml with water. Dispense from dropper bottles. These 0.02 M solutions of the amino acids in approximately 0.02 M HCl are stable for long periods when stored in the refrigerator.

The amino acids should be weighed on an analytical balance. The quantity of each (0.001 mole) to prepare 50 ml of 0.02 M solutions are given in the table below.

	grams		*grams*
alanine	0.089	lysine·HCl	0.183
arginine·HCl	0.211	methionine	0.149
aspartic acid	0.133	phenylalanine	0.165
glutamic acid	0.147	proline	0.115
glycine	0.075	serine	0.105
histidine·HCl·H$_2$O	0.210	tryptophan	0.204
isoleucine	0.131	tyrosine	0.181
leucine	0.131	valine	0.117

— — *Unknown amino acid solutions.* Three drops of each of three amino acids from the student's assigned group in a 75 mm test tube. See the various groups of amino acids listed on page 177.

Experiment 26. Precipitation of Proteins

5-6 liters *Egg white (egg albumin),* 1% solution. See Expt. 24, p. 286.

500 ml *Urea,* 1% solution. Dissolve 5 g urea in 500 ml water.

300 ml *Alanine,* 1% solution. See Expt. 24, p. 285.

A. Concentrated acids

600 ml each *Sulfuric acid, Nitric acid,* and *Hydrochloric acid.* Dispense from special acid burets, see p. 268.

B. Heavy metal ions. Dispense from dropper bottles.

50 ml *Copper sulfate,* 0.2 M solution. Dissolve 5.0 g CuSO$_4$·5H$_2$O in 100 ml water.

50 ml *Lead acetate,* 0.2 M solution. Dissolve 7.6 g Pb(C$_2$H$_3$O$_2$)$_2$·3H$_2$O in 100 ml water.

50 ml *Mercuric chloride,* 0.2 M solution. Dissolve 5.4 g HgCl$_2$ in 100 ml water.

C. Alkaloidal reagents. Dispense from dropper bottles.

50 ml *Phosphotungstic acid,* 20% solution. Dissolve 20 g P$_2$O$_5$·24WO$_3$·nH$_2$O in about 50 ml water and dilute to 100 ml.

50 ml *Tannic acid,* 5% solution. Dissolve 5 g tannic acid in about 75 ml water and dilute to 100 ml.

50 ml *Trichloroacetic acid (TCA),* 10% solution. Dissolve 10 g CCl$_3$COOH in water and dilute to 100 ml.

D. Salting-out of egg proteins

1000 g *Ammonium sulfate,* solid.

300 ml *Sodium hydroxide,* 10% solution. See Expt. 2, p. 271.

50 ml *Copper sulfate,* 0.1% solution. Dispense from dropper bottle. See Expt. 2, p. 270.

100 *Filter paper,* 11 cm, Whatman No. 40.

E. and F. Organic solvent and Sevag test

300 ml *Acetone.*

300 ml *Ethyl alcohol,* 95%.

100 ml *Chloroform.*

Experiment 27. Kjeldahl Semimicrodetermination of Nitrogen and Crude Protein

1-2	*Kjeldahl Digestion Apparatus.* Rotary 12-place digestion apparatus. (Kontes catalog No. K-551000-0000 with fume hood cat. No. K-551001-0025; or American Instrument Co. catalog No. 4-2460 with fume hood cat. No. 4-1810. Other types of digestion apparatus for 100-ml Kjeldahl flasks can also be used.
2-4	*Kjeldahl distillation rack,* twin unit, rheostat controlled. Made by Precision Scientific Co. (No. 55126); available from Scientific Products, Sargent-Welch, and Fisher Scientific. Less expensive assemblies can be improvised with glass condenser and portable Precision electric heaters for Kjeldahl flasks.
120-150	*Flasks, Kjeldahl,* 100-ml, long neck.
6-12	*Traps,* connecting, Kjeldahl, Iowa State-type. Other types, suitable for 100-ml Kjeldahl flasks, can be used.
2 pkg.	*Cigaret paper.* Available at most drug stores.
200	*Glass beads.*
2 liters	*Boric acid,* 2% solution. Dissolve 20 g boric acid in water to make 1 liter of solution.
325 ml	*Mercuric sulfate solution.* Carefully add 36 ml conc. H_2SO_4 to 300 ml water. Dissolve 30 g red mercuric oxide in this solution. Dispense from a 50-ml buret.
50 ml	*Methyl red-methylene blue indicator (Tashiro's indicator).* See Expt. 8, p. 275.
325 g	*Potassium sulfate,* solid.
2 liters	*Sodium hydroxide (13 N)/sodium thiosulfate (5%) solution.* Gradually add, with constant stirring, 530 g NaOH pellets (98% assay) or NaOH technical flakes (For N determination) to 750 ml water using a 2-liter beaker. Considerable heat is evolved and cooling of the beaker in a cold water bath may be needed. *Constant stirring is important to avoid solidification* of NaOH on bottom of beaker. After cooling the solution (cover with watch glass), add 50 g $Na_2S_2O_3 \cdot 5H_2O$ and dilute to 1 liter. Store the caustic solution in a polyethylene bottle. Keep tightly stoppered to prevent absorption of carbon dioxide.
600 ml	*Sulfuric acid,* concentrated. Dispense from special acid buret, see p. 268.
4-5 liters	*Sulfuric acid,* 0.05 N solution (for titrating distilled ammonia). Pipet exactly 50.0 ml of 1 N H_2SO_4 (see p. 269) into a 1-liter volumetric flask and dilute to the mark. To prevent waste of reagent, set up several 50-ml burets for titrating NH_3 distillates.
	Unknown samples. A series of analyzed dried blood samples ranging from approximately 2 to 14% nitrogen can be purchased from Smith & Underwood Laboratories, 320 E. Fourth St., Royal Oak, Michigan 48067. About 25 different N values are available in 10 g vials and ¼ lb. bottles. Each student can be given 0.5-1 gram of unknown in a 75-mm test tube.

Experiment 28. Biuret Determination of Protein

5-10	*Spectrophotometers.* Bausch and Lomb Spectronic 20 or similar type.
3.2 liters	*Biuret reagent.* Transfer 1.5 g $CuSO_4 \cdot 5H_2O$ and 6.0 g of sodium potassium tartrate tetrahydrate (Rochelle salt) into a 1-liter volumetric flask. Add 500 ml of water and 300 ml carbonate-free 10% NaOH solution and mix. Dilute to 1 liter with water and store in a polyethylene bottle. Dispense from 50-ml buret.
2 liters	*Sodium chloride,* 1% solution. Dissolve 10 g NaCl in water and make to 1 liter.

750-1000 ml	*Standard protein solution,* 10 mg protein/ml. Prepare the volume of protein solution needed in an appropriate size volumetric flask at a concentration of exactly 1.00 g bovine serum albumin (Fraction V), or egg white lysozyme, per 100 ml 1% NaCl solution. Store in refrigerator. (Bovine serum albumin (Fraction V) or egg white lysozyme are available from Sigma Chemical Co. or other suppliers.)
— —	*Unknown protein solutions.* A series of unknowns can be prepared by diluting aliquots of the above standard protein solution *with 1% NaCl* in 50 or 100-ml volumetric flasks (depending on number of students) and giving each student 4-5 ml of an unknown solution.

Experiment 29. Electrophoresis of Serum Proteins

	For this experiment the equipment used is made by Gelman Instrument Co., Ann Arbor, Michigan 48106, and is available through Scientific Products and other suppliers. Similar types of equipment can be used.
1-2	*Electrophoresis chamber* (Gelman 51211).
1-2	*Power supply* (Gelman 38206).
box of 100	*Electrophoresis strips,* cellulose polyacetate, Sepraphore III, 2.54 x 15.2 cm (1″ x 6″). (Gelman 51003). Soak strips in pH 8.8 buffer in one of the "staining trays." For more effective wetting of immersed strips, add 1 ml ethyl alcohol per 100 ml buffer. Strips must be immersed face-up as in its box. Store in refrigerator.
6-8 (50/pkg.)	*Absorbent pads,* 7.5 x 15 cm (3″ x 6″), (Gelman 62093). Draw a vertical pencil line 5 cm (2″) from the left end of a pad for a guide during sample application.
100	*Capillary pipet,* disposable, 10 μl. Dade "Accupette," (Scientific Products, catalog No. P4518-10). For transferring serum sample to wires of sample applicator. For convenient handling, pipet can be inserted into the "Microcap" pipet holder used in Expts. 21 and 25.
3-6	*Forceps,* plastic.
6-8	*Glass plates,* chromatographic, 10 x 20 cm (4″ x 8″). For clearing and drying electrophoresis strips.
24-36	*Magna-Grip Tensioners* (Gelman 51139).
2	*Sample applicator* (Gelman 51225).
12	*Staining trays* (Gelman 51159). These multi-purpose trays are used for soaking, staining, rinsing, and dehydrating the electrophoresis strips.
1-2	*Tray,* glass, for clearing strips. A Pyrex baking pan (houseware department of stores) approximately 6″ x 10″ x 1½″ deep is suitable for clearing strips with the 13% acetic acid in methanol. Cover the tray with aluminum foil when not in use to prevent evaporation of the reagent.
1 roll	*Transparent tape.* For mounting cleared electrophoresis strip on report sheet.
— —	*Blood serum sample.* A blood serum sample of 2-3 ml, if kept refrigerated when not in use, is stable for about a week. Alternatively, an "Electrophoresis Control" (Gelman 51913) is available in 6 vials per box which is sufficient for 3-4 semesters. The lyophilized proteins are reconstituted with 2 ml of water as needed.
4-5 liters	*Acetic acid,* 5% solution. Add 50 ml of glacial acetic acid to water and dilute to 1 liter.
1 liter	*Acetic acid-methanol clearing solution,* 13% CH_3COOH in CH_3OH. Mix 13 volumes of glacial acetic acid with 87 volumes of anhydrous methanol.

2400 ml (2 x 18 g vials)	*Buffer solution*, pH 8.8 (Gelman 51104). Dissolve 18 g (1 vial) of Tris/barbital buffer in 1200 ml water. Stable in refrigerator for 1-2 weeks. Can use 4-5 times.
1 liter	*Methanol*, anhydrous. For dehydrating electrophoresis strips after rinsing stained strips in 5% acetic acid.
200 ml	*Ponceau S dye*, 0.5% dye in 5% TCA. Dissolve 0.5 g dye in 100 ml of 7.5% aqueous trichloroacetic acid (TCA) solution. Stable in refrigerator for 1-2 weeks. The Ponceau S dye powder is available from most chemical suppliers.
2 liters	*Sodium hydroxide*, 0.1 N solution. See p. 269. Not required unless part F (elution) is done.

Experiment 30. Colorimetric Determination of Ascorbic Acid (Vitamin C)

	NOTE: The instructor will determine which fruits, fruit juices, or vegetables will be assigned to the students. These materials can be furnished by the laboratory, or the students can be requested to bring an adequate amount of their assigned material to the laboratory.
1	*Blender*, Waring or similar type.
10-20	*Burets*, 10-ml with 0.05 ml divisions.
— —	*Burets*, 50-ml. If not available as locker equipment, 20-25 burets are needed from the stockroom or reagent shelf.
100	*Filter paper*, 11 cm, medium grade.
bottle	*Glass wool.*
1-2 liters	*2,6-Dichloroindophenol dye solution.* Dissolve 0.5 g of 2,6-dichloroindophenol sodium salt in 975 ml water. Add 25 ml of 0.05 M phosphate buffer pH 7.0. Dispense from several 10-ml burets on reagent bench. To fill burets, use a 500-ml "squeeze bottle" containing the dye solution. (**NOTE:** For pH 7 buffer, dilute the 0.1 M pH 7 buffer used in Expt. 32, p. 293 or dilute and adjust to pH 7 with NaOH the 0.1 M dihydrogen phosphate buffer used in Expt. 7, p. 274.)
6-10 liters	*Oxalic acid*, 1% solution. Prepare at a concentration of 10 g oxalic acid dihydrate per 1 liter of water.
2.5 liters	*Ascorbic acid standard solution*, 0.1 mg/ml. Accurately weigh 50 mg of ascorbic acid, transfer quantitatively to a 500-ml volumetric flask, and dissolve in 500 ml *1% oxalic acid solution.* Prepare a fresh solution daily.

Experiment 31 Extraction and Chromatography of Carotenes. Provitamin A

	NOTE: The volumes of reagents listed below for parts A and B are for the preparation of *one extract* of carotenes as described in the procedure. This is sufficient for a demonstration of column chromatography only. If greater student participation is desired, the instructor will need to provide for the preparation of additional extracts of carotenes. All students can do the thin-layer chromatography of an extract.

Extraction and Column Chromatography of Carotenes (parts A and B)

20 g	*Carrots.*
1-2	*Chromatographic column*, 11 mm i.d. x 300 mm length, with removable Teflon stopcock which also fits student 25 and 50-ml burets. (Kimax No. 17800 and Pyrex No. 2145) Columns also available without stopcock if you use the stopcock assemblies from the student burets. A 25-mm buret can be used for a chromatographic column if necessary.

1	*Glass rod,* 10 mm o.d. x 35 cm long, for packing column.
bottle	*Glass wool.*
1	*Grater,* household vegetable type.
1	*Separatory funnel,* 500 ml.
200 ml	*Ethyl alcohol,* 95%.
150 ml	*Petroleum ether,* (Skellysolve B, B.P. 60-70°). See Expt. 21, p. 284.
10 g	*Magnesium oxide-Celite adsorbent.* Intimately mix equal weights of MgO powder (suitable for chromatographic use) and Celite. A 1:1 mixture of MgO-Celite is available from Merck & Co.

Thin-Layer Chromatography of Carotenes (part C)

15	*Chromatography sheets,* precoated silica gel, 20 x 20 cm. Cut each 20 x 20 cm sheet into eight 5.0 x 9.5 cm (2″ x 3¾″) sheets. Activate in oven at 100° for 15-30 minutes. Store in a desiccator. See Expt. 21, p. 284.
100	*Micropipets,* one microliter (1 μl). See Expt. 21, p. 284.
– –	*Carotene extract,* see part A above.
10 ml	*β-Carotene reference,* 0.05% solution in petroleum ether. Dissolve 0.05 g carotene in 100 ml petroleum ether. Provide several small vials of the solution for student application. (Sigma Chemical Co., St. Louis, Mo. Catalog No. C9875, Type III, 80-90% β-carotene and 10-20% α-carotene.)
1 liter	*Solvent system,* for developing chromatogram. Suggested solvent systems: (1) Petroleum ether, benzene (50:50 v/v). **CAUTION:** Benzene vapors are toxic, use under hood. (2) Petroleum ether, ethyl ether, glacial acetic acid (70:30:1, v/v). See Expt. 21, p. 284. These solvent systems are flammable; keep away from flames!
100 ml	*Spray solution,* 5% paraffin oil in petroleum ether. (Optional) Dissolve 5 ml paraffin oil in 100 ml petroleum ether in a laboratory sprayer. Use in good fume hood away from flames, hot plates, etc. Vapors highly flammable and potentially hazardous!

Experiment 32. Enzymatic Digestion of Proteins. Factors Affecting Enzyme Activity

3-4 rolls	*Photographic film negative.* Expose an unrolled photographic film (Verichrome Pan 120 or similar type) to bright light for 1-2 minutes. The film is developed, fixed, washed and dried. The black negative is cut into 8 x 12 mm (5/16″ x ½″) pieces for student use.

The exposed film can be processed as indicated below. All processing steps can be done in a lighted room since you are not making pictures. If you use a film tank, keep the film in the tank until after the washing step.

(1) *Development.* Use Kodak D-76 or Kodak Microdol-X developer. Prepare and use developer in accordance with manufacturer's recommendations. Developing time: D-76 use 16 minutes (tray or beaker) or 20 minutes (tank); Microdol-X use 9 min. (tray or beaker) or 13 minutes (tank). Developer can be used again, but increase developing time slightly with each roll.

(2) *Acid short-stop.* Pour the developer solution into its storage bottle. Rinse film in short-stop solution (13 ml glacial acetic acid in 900-1000 ml water). If film tank is used, add acid solution to cover film. Agitate few times and discard acid solution in sink. Do *not* leave film in acid solution for more than a minute.

(3) *Fixing.* Prepare a solution of Kodak Fixer in accordance with manufacturer's directions. Fix the black negative in a large beaker or in film tank for 15-20 minutes. Agitate frequently. Return fixing solution to storage bottle.

(4) *Washing.* Wash film in water for 15-20 minutes using *frequent* changes of water. *Water must not exceed 25°C;* warm water will damage the gelatin on the film. Slowly running water can be used by running rubber tubing to *bottom* of beaker or film tank allowing it to overflow.

(5) *Drying and Cutting.* Allow film to air-dry by hanging it with clips from a tall ring stand. After drying, cut film into 8 x 12 mm pieces and store in a small wide-mouth bottle. Approximately 500 pieces should be obtained from 1 roll of Verichrome Pan film.

2 rolls   *Transparent tape.* For mounting film strips on report sheet.

1-2   *Water bath,* laboratory type with plastic test tube racks, at 40°C. If large bath is not available, students can use a 600-ml beaker on a ring stand, *carefully* controlling the water temperature with burner and thermometer.

3.5 liters   *Pancreatin extract.* Each student uses 35 ml pancreatin solution for the entire experiment. Prepare only sufficient solution for one day's use; discard unused solution. The concentration of the freshly-prepared solution is 0.13 g pancreatin powder per 100 ml of water (see NOTE below). Mix the desired volume of pancreatin solution at a moderate rate on a magnetic stirrer for at least 20-30 minutes. Dispense from two 50-ml burets on reagent bench. Fill burets from a *labeled* polyethylene "squeeze bottle." Store in refrigerator when not in use. *Make fresh solution when needed.*

IMPORTANT NOTE: When 3 ml of pancreatin extract solution are used with 3 ml pH 8 buffer at 40°C, the digestion time to clear a film strip should be *not less than 15 minutes.* The 0.13 g pancreatin powder per 100 ml water given above is based on a certain batch of powder from Fisher Scientific. Depending on the source of pancreatin, larger or smaller amounts of powder may be necessary to give the minimum 15-minute digestion time. If digestion time is less than 15 minutes, the factors affecting enzyme activity can not be properly observed.

3.5 liters   *Formaldehyde solution,* 18-19% HCHO. Mix 37-38% formaldehyde solution with an equal volume of water (1:1 v/v). Place 500-ml bottles of solution in the ventilating hood. Irritating to eyes.

— —   *Ice,* crushed or chipped. For ice bath in part C.

3 liters   *Phosphate buffer, pH 8,* 0.1 M solution. Dissolve 13.8 g $NaH_2PO_4 \cdot H_2O$ in about 900 ml of water and add 1 M NaOH slowly with mixing (magnetic stirrer) until the pH reaches 8.0 (pH meter). Dilute to 1 liter.

300 ml   *Phosphate buffer, pH 5,* 0.1 M solution. Dissolve 6.9 g $NaH_2PO_4 \cdot H_2O$ in about 400 ml water and slowly add 1 M NaOH until the pH reaches 5.0 (pH meter). Dilute to 500 ml.

300 ml   *Phosphate buffer, pH 6,* 0.1 M solution. Proceed as with the pH 5 buffer above, but add 1 M NaOH to give pH 6.0.

300 ml   *Phosphate buffer, pH 7,* 0.1 M solution. Proceed as above, but add 1 M NaOH to reach pH 7.0.

NOTE: Dispense all buffer solutions from 50-ml burets on reagent bench to prevent waste.

15 ml   *Mercuric chloride,* 0.2 M solution. Dispense from a dropper bottle. See Expt. 26, p. 288.

Experiment 33. Hydrolysis of Starch by Salivary Amylase

20	*Spot plate,* porcelain. Usually part of locker equipment, or to be checked out of stockroom.
500 ml	*Barfoed's reagent.* See Expt. 10, p. 276.
500 ml	*Benedict's reagent.* See Expt. 10, p. 276.
150 ml	*Iodine,* 0.02 M solution. Dissolve 0.5 g iodine and 2 g potassium iodide in 150 ml water. Dispense from dropper bottles.
1 liter	*Starch,* 1% solution. See Expt. 10, p. 276.

Experiment 34. Enzymatic Digestion of Fat

A. Milk Fat Digestion

1-2	*Water bath.* See Expt. 32, p. 293.
1 liter	*Milk,* whole or 2%.
50 ml	*Litmus solution.* Prepare 250 ml of a saturated litmus solution and filter. Dispense from dropper bottles.
	Pancreatin solutions:
300 ml	(1) *Unboiled.* Disperse 4 g of pancreatin powder in 900 ml of 50% alcohol (magnetic stirrer) and adjust to pH 8.0 with 0.1 M NaOH using a pH meter. Dilute to 1 liter with water. It is important that the pH is 8.0 so that the effect of liberated fatty acids can be properly noted on the litmus color. Stable for one week if stored in the refrigerator *when not in use.*
300 ml	(2) *Boiled.* Under reflux, boil 500 ml of the pancreatin solution prepared in (1) above. (Do not use an open beaker with a flame.) After refluxing about 5-10 minutes, cool and store in the refrigerator.
10 ml	*Sodium carbonate,* 10% solution. Dissolve 10 g anhydrous Na_2CO_3 in water to make 100 ml of solution. Dispense from dropper bottles.

B. Effect of Bile Salts on Lipase Action

3 liters	*Pancreatin solution.* Prepare as described above in (1) for "unboiled" pancreatin solution. Refrigerate when not in use.
100 ml	*Phenolphthalein indicator,* 1% solution. See Expt. 8, p. 275.
300 ml	*Sodium cholate,* 5% solution. Suspend 25 g of cholic acid in 400 ml water using a magnetic stirrer. Add 10% NaOH solution slowly until the pH reaches 8.0 (pH meter) and the solid has dissolved. Dilute to 500 ml with water.
4-5 liters	*Sodium hydroxide,* 0.02 N solution. Dilute exactly 200 ml 0.100 N NaOH (p. 269), or 20.0 ml of 1 N NaOH, to the mark with water in a 1-liter volumetric flask.
200 ml	*Vegetable oil,* corn or cottonseed. Recommend dispensing from a 50-ml buret on reagent bench so that many student pipets are not messed up.

Experiment 35. Fermentation of Sugars

24	*Fermentation tubes,* Smith, closed tube, 13 mm i.d. x 130 mm in length (Pyrex No. 9460). If different size of tube is used, see part A of Procedure for adjusting amounts of yeast and sugar. If large tubes are used (17 x 150 mm), support them across top of a 800-ml beaker.
100	*Filter paper,* 11 cm, Whatman's No. 41.
bottle	*Glass wool.*

1 liter	*Glucose,* 5% solution (part C). Dissolve 25 g glucose in water and make to 500 ml.
bottle	*Iodine,* crystals.
500 ml	*Sodium hydroxide,* 5% solution. Dissolve 25 g NaOH in about 400 ml of water with constant mixing and dilute to 500 ml.
250 ml of each	*Sugars,* 2% solution. Using 2 g of the sugar per 100 ml water, prepare separate solutions of the following: fructose, glucose, galactose, xylose, lactose, maltose, and sucrose.
250 ml	*Starch,* 2% solution. Make a thin slurry of 5 g of soluble starch and pour it into 250 ml of boiling water. Boil 1 minute and cool.
100-150 g	*Yeast, active dry,* baker's or brewer's. Available from Sigma Chem. Co., St. Louis, MO 63178, and other sources. If compressed yeast is substituted, use about three times the amount specified for dry yeast.

Experiment 36. Bioluminescence. Determination of ATP

— —	*Yeast* (compressed), muscle or kidney tissue. See part A, page 241, for further details.
100 ml	*ATP solution,* 0.001 (1×10^{-3}) M solution. Dissolve 55.1 mg adenosine-5'-triphosphate disodium salt in water and dilute to 100 ml. Prepare fresh and keep refrigerated. Stable for several months if kept frozen. ATP available from Sigma Chemical Co., St. Louis, Mo. 63178.
10 ml of each	*Diluted ATP solutions.* A series of diluted ATP solutions are prepared by mixing the volumes of 1×10^{-3} M ATP and water given in the table below to give 10 ml of the diluted solution.

ATP molarity	1×10^{-3} *M ATP*	*water*
5×10^{-4}	5.0 ml	5.0 ml
2×10^{-4}	2.0 ml	8.0 ml
1×10^{-4}	1.0 ml	9.0 ml
5×10^{-5}	0.5 ml	9.5 ml

100 ml	*Arsenate buffer,* pH 7.4, 0.1 M solution. Dissolve 3.12 g $Na_2HAsO_4 \cdot 7H_2O$ in about 80 ml water and adjust to pH 7.4 with dilute HCl (pH meter). Dilute to 100 ml.
100 ml	*Glycine buffer,* pH 7.4, 0.1 M solution. Dissolve 0.375 g glycine in 80 ml water and adjust to pH 7.4 (pH meter). Dilute to 100 ml.
100 ml	*Magnesium sulfate,* 0.1 M solution. Dissolve 2.46 g $MgSO_4 \cdot 7H_2O$ in water and dilute to 100 ml.
100 ml	*Trichloroacetic acid (TCA),* 5% solution. Dissolve 5 g trichloroacetic acid in 100 ml water.
— —	*Firefly extract.* Desiccated firefly lanterns available from Sigma Chemical Co. These are ground in a mortar with an equal weight of sand and the arsenate buffer in the proportion of 100 mg desiccated lanterns to 10 ml arsenate buffer. The buffer is used in three portions and is decanted into a centrifuge tube after each grinding. The combined liquid is then placed in a deep-freeze cabinet until frozen, then allowed to thaw, and the inactive precipitate centrifuged out. The somewhat cloudy supernatant contains luciferin and luciferase. Prepare immediately before use and keep in ice. The enzyme is stable for several months, however, if kept frozen.

— —   *Firefly reagent mixture,* for assay of ATP. Mix the following solutions immediately before the laboratory period:

        5.0 ml freshly-prepared firefly lanterns in 0.1 M arsenate buffer. Do NOT pipet by mouth (poison!)

        3.5 ml 0.1 M magnesium sulfate solution

        3.5 ml glycine buffer, pH 7.4

Experiment 37.   Cytochrome *c* Oxidase (Cytochrome *a a*₃)

1500 ml   *Phosphate buffer,* pH 7.3, 0.05 M solution. Dissolve 6.9 g $NaH_2PO_4 \cdot H_2O$ in 500 ml of water and add 0.1 M NaOH (about 400 ml) until the pH is 7.3 *(use pH meter).* Dilute to 1 liter.

150 ml   *Yeast, stock suspension.* Suspend 50 g of compressed yeast (not dry yeast) in 100 ml of phosphate buffer (pH 7.3). Add 1 ml of 10% Triton X-100 solution and mix. The surfactant increases yeast cell permeability.

100 ml   *Yeast, heated stock suspension.* Rapidly heat a portion (100 ml per 100 students) of the stock solution prepared above to 55-57° with *constant* stirring and then place it in a 57° ± 1° water bath for 1 hour. The temperature of the water bath must be kept within one degree. Keep stock yeast suspensions in refrigerator when not in use.

— —   *Yeast, dilute unheated suspension.* Prepare a 10-fold dilution at the beginning of the laboratory period. For 20-25 students, dilute 5 ml of *unheated* stock yeast suspension with 45 ml of *phosphate buffer* (pH 7.3). Clearly label "UN-HEATED." Do not keep unused diluted suspensions.

— —   *Yeast, dilute heated suspension.* Make a 10-fold dilution at the beginning of the laboratory period. For 20 students, dilute 20 ml of the *heated* (at 57°C, see above) stock yeast suspension with 180 ml of *phosphate buffer* (pH 7.3). Clearly label "HEATED at 57°." The yeast suspension can be readily dispensed from a 25-ml buret if used within 15-20 minutes after filling buret, or use other suitable dispenser. Discard unused diluted yeast suspension.

125 ml   *p-Aminodimethylaniline,* 0.01 M solution. Dissolve 0.172 g of p-aminodimethylaniline hydrochloride (dimethyl-paraphenylenediamine·HCl) in 100 ml of 50% ethanol.

125 ml   *α-Naphthol,* 0.01 M solution. Dissolve 0.144 g of α-naphthol in 100 ml 50% ethanol.

125 ml   *Sodium carbonate,* 0.1% solution. Dissolve 0.1 g anhydrous Na_2CO_3 (or 0.12 g $Na_2CO_3 \cdot H_2O$) in 100 ml of water.

375 ml   *Nadi reagent.* Mix equal volumes of the following: 0.01 M p-aminodimethylaniline, 0.01 M α-naphthol, and 0.1% sodium carbonate. For 20 students, prepare 75 ml of the mixture immediately before use and dispense from a 25-ml buret. Discard unused reagent.

25 ml   *Potassium cyanide,* 0.01 M solution. Dissolve 0.07 g KCN in 100 ml of water. Dispense from dropper bottle.

100 ml   *Hydrogen sulfide,* saturated (0.1 M) solution. (1) Bubble H_2S gas from a suitable generator (in hood) to saturate 100 ml of water. "Aitch-tu-ess" cartridges (Fisher Scientific Co.) can be used; or (2) dissolve 2.4 g sodium sulfide ($Na_2S \cdot 9H_2O$) in 80 ml of water and add 20 ml 1 N HCl solution (hood). Dispense in a dropper bottle.

Experiment 38. Urease. Determination of Urea in Blood

> **NOTE:** Do *not* use acetone to dry glassware. Ammonia-free, deionized water is preferable for preparing reagents to reduce blank absorbance. Reagents should be dispensed from burets to prevent waste and for convenience.

1-2   *Centrifuge.* If not available, tungstate-treated samples can be clarified by filtration through dry filter paper.

— —   *Ammonium sulfate,* 0.100 N stock solution. Dissolve 1.651 g of pure $(NH_4)_2SO_4$ in water in a 250-ml volumetric flask and dilute to exactly 250 ml.

200 ml   *Ammonium sulfate,* 0.001 N standard solution. Pipet 1.00 ml of 0.100 N $(NH_4)_2SO_4$ stock solution into a 100-ml volumetric flask and dilute to the mark. Dispense from a *10-ml buret.*

800 ml   *Nessler's reagent.* Dissolve 4 g KI and 4 g HgI_2 in about 20 ml of water. Add 1.75 g gum ghatti and dilute to 1 liter. Dispense from a 50-ml buret.

1200 ml   *Sodium hydroxide,* 2 N. With *constant* stirring, dissolve 80 g of NaOH pellets in about 600 ml of water, cool and dilute to one liter. Dispense from a 50-ml buret.

200 ml   *Sodium tungstate,* 10% solution. Dissolve 10 g $Na_2WO_4 \cdot 2H_2O$ in water and dilute to 100 ml. Dispense from a 25-ml buret.

300 ml   *Sulfuric acid,* 1 N solution. See page 269. Dispense from a 25-ml buret.

200 ml   *Urease solution,* 1.6 Sumner units per ml. Dissolve that amount of jack bean powder that contains 160 Sumner units of urease in 100 ml of 30% ethanol. (Consult reagent bottle label for activity. Sigma Chemical Co., St. Louis, Mo. 63178 has a Type IV jack bean powder with a urease activity of 2,000-4,000 units per gram.)

(1) *Active Urease solution.* Use one-half of the urease solution prepared as described above. Dispense from a *10-ml* buret.

Clearly label the buret: "ACTIVE UREASE."

(2) *Inactive (Boiled) Urease solution.* Boil the remaining half of the urease solution for about 5 minutes in a beaker covered with a watch glass. Replace any loss in volume with water. Dispense from a *10-ml* buret.

Clearly label the buret: "INACTIVE UREASE—BOILED."

Experiment 39. Preparation and Properties of Nucleic Acids

A. Homogenization and Extraction of Rat Liver

— —   *Rats,* Approx. 6-10 g liver per rat depending on size. One rat should give sufficient homogenate for 15 to 25 students. See procedure.

1   *Blender,* Waring or other model.

1-2   *Centrifuge,* International clinical or other suitable model for use with 15-ml conical centrifuge tubes.

24   *Centrifuge tubes,* 15-ml, conical.

1 liter   *Ethanol,* 95%.

130 ml   *Trichloroacetic acid (TCA),* 10% solution. Dissolve 50 g CCl_3COOH in water and make to 500 ml.

250 ml   *Trichloroacetic acid,* 5% solution. Dilute 10% TCA with equal volume of water.

750 ml   *Trichloroacetic acid,* 2.5% solution. Dilute 1 volume of 10% TCA with 3 volumes of water.

B. Color Tests on Nucleic Acid Components

250 ml   *Ribonucleic acid,* 0.02% RNA solution in 5% TCA. Dissolve 0.05 g commercial RNA in 250 ml hot 5% TCA (do not exceed 90°). Dispense from 25-ml buret. (RNA available from Sigma Chemical Co., St. Louis, Mo. 63178, or other source).

150 ml   *Deoxyribonucleic acid,* 0.05% DNA solution in 5% TCA. Dissolve 0.10 g commercial DNA in 200 ml hot 5% TCA (do not exceed 90°). Dispense from 25-ml buret. (DNA available from Sigma Chemical Co. or other source).

150 ml   *Ribose* or *Xylose,* 0.01% solution. Dissolve 0.02 g ribose or xylose in 200 ml water, or dilute 2 ml of 1% ribose or xylose (Expt. 10, p. 276) to 200 ml with water.

150 ml   *Glucose,* 0.01% solution. Dissolve 0.02 g glucose in 200 ml of water, or dilute 2 ml of 1% glucose to 200 ml with water. (Expt. 10, p. 276).

200 ml   *Ammonium molybdate,* 2.5% solution. Dissolve 5 g $(NH_4)_6Mo_7O_{24} \cdot 4H_2O$ in 200 ml of water.

small btl.   *Barium hydroxide,* $Ba(OH)_2 \cdot 8H_2O$. Keep bottle tightly closed. Absorbs CO_2 readily.

100 ml   *Bromine water,* saturated with liquid Br_2. Carefully add small amounts of liquid bromine (CAUTION, USE HOOD) to 100 ml of water until saturated. Dispense from dropper bottle.

1 liter   *Diphenylamine reagent.* Dissolve 2 g of pure diphenylamine in 200 ml glacial acetic acid and 5.5 ml conc. H_2SO_4. The reagent is stable on storage. Dispense from 50-ml buret on reagent bench. Caution students in its use.

10 ml   *Nitric acid,* 2 N solution. Add 12.6 ml conc. HNO_3 to water to make 100 ml of solution. Dispense in dropper bottle.

1 liter   *Orcinol reagent.* See Expt. 11, p. 277. Dispense from 50-ml buret on reagent bench.

50 ml   *Potassium hydroxide,* 10% solution. Dissolve 10 g KOH pellets in water with constant stirring and dilute to 100 ml. Dispense from dropper bottle without ground-glass joint.

5 g ea.   *Purine bases: Adenine and guanine. Also uric acid.* A student requires only a few mg of each.

250 ml   *Pyrimidine base: Cytosine or uracil,* saturated solution. Dissolve 0.77 g cytosine or 0.36 g of uracil in water. Warm if necessary.

200 ml   *Reducing reagent* (Modified Fiske-SubbaRow reagent). See reagents for Expt. 5, p. 273. Dispense from dropping bottle. Stable 10 days in brown bottle.

100 ml   *Sodium dihydrogen phosphate,* 0.001 M solution. Dilute 1.0 ml of 0.1 M NaH_2PO_4 (Expt. 7, p. 274) to 100 ml with water.

200 ml   *Sulfuric acid,* 5 N solution. Cautiously and slowly add 28 ml of conc. H_2SO_4 to 150 ml of distilled water. Cool and dilute to 200 ml. Dispense from dropper bottle.

Index

Moisture in biological materials
  content of, 2
  detection of, 2, 5, 6
  quantitative determination of, 2–6
Molarity of acids and bases, 268
Molisch test, for soluble carbohydrates, 66–67
Monoacylglycerols, from enzymatic digestion of fats, 231
Monochromatic light
  definition of, 17
  production of, 18, 19
  transmission through solutions, 17
Monoglycerides. See Monoacylglycerols
Monosaccharides, distinguishing from other carbohydrates using Barfoed's test, 71, 72
Mucic acid (galactaric acid), from strong oxidation of galactose, 71
Mucic acid test, for galactose and galactose-containing carbohydrates, 71, 72
Murexide test, for purines and pyrimidines, 255
Nadi reagent
  in detection of cytochrome c oxidase, 245–248
  preparation and use of, 246
  reactions in presence of cytochrome c oxidase, 245
Nessler reagent
  composition of, 249
  preparation of, 297
  sensitivity to acetone, 249
Nicotinamide adenine dinucleotide (NAD⁺), 235
Ninhydrin
  reaction with amino acids, 169
  structure of, 171
Ninhydrin reagent, for paper chromatography, 176, 287
Ninhydrin test
  for amino acids and proteins, 169, 170
  response to, by various substances, 169
Nitric acid, concentrated
  concentration values of, 268
  corrosiveness of, 72
  dispensing buret for, 268
  oxidation of sugars by, 71, 72
  xanthoproteic test using, 169, 170
Nitric acid, dilute solutions, 268, 269
Nitrogen in proteins
  Kjeldahl determination of, 189–192
  test for, 163
Nonmetals
  loss during dry ashing of biological materials, 11
  nutritionally essential, 11
Nonsaponifiable materials, 125–128
  definition of, 125
  separation from polar compounds, 125
Normal solutions
  of acids and bases, 57
  standardization of, 269
Normality (N)
  definition of, 57
  units for, 57
Nucleic acids
  color tests for components of, 255–256
  hydrolysis products of, 254
  preparation from liver, 254
  purification by dialysis of, 7
  separation from other components of tissue, 254

Oil of wintergreen, methyl salicylate in, 113
Optical density. See Absorbance
Optical system of Spectronic 20 colorimeter, 19
Orcinol, chemical name of, 77
Orcinol reagent, preparation of, 277
Orcinol test
  comparison with aniline acetate test, 77
  for pentoses and pentosans, 77, 78
  for ribose in nucleic acids, 255
Organic solvents
  fat extraction with, 155
  precipitation of proteins by, 183, 184
Orthophenanthroline colorimetric method for iron, 17–30, 33, 35
Overtitration, 58
Oxalated blood plasma, preparation of, 96
Oxidases, 245
Oxidizing agents, for ashing of biological materials, 11, 31
Pancreatic extracts
  for hydrolysis of fats, 231
  for hydrolysis of gelatin on photographic film, 221
Pancreatin powder, 221, 231
Paper, chromatographic,
  cutting of, 175
  grain of, 175
Paper chromatography, 175–182
  advantages of, 175
  of amino acids, 175–182
  solvent systems for, 175, 177
  two-dimensional, 175
Paper, glassine, for weighing samples, xiv, 59, 60
Partition coefficient, relation to chromatography, 147, 148
Pentosans
  hydrolysis to pentoses, 66
  production of furfural from, 66, 77
Pentoses
  color test for, 77–82
  production of furfural from, 66, 77
Peptide bonds
  chemical nature of, 169
  hydrolysis of, 221
  minimum number of, in proteins, 169
  requirement for, in biuret test, 169
Peroxidase, catalyst for hydrogen peroxide decomposition, 95, 129, 130
pH
  control of, by buffers, 49
  measurement of, 42–48
    by indicator paper, 42–48
    by pH meter, 42, 43
  of prepared buffer solutions, 51–52
pH calculations
  from hydroxide ion concentration, 42
  from Henderson-Hasselbalch equation, 51
pH meters
  description of, 43, 44
  illustrations of, 44
  standardization of, 42
  use of, 43
  various types of, 43, 44
Phenanthroline. See Orthophenanthroline
Phenolphthalein indicator
  preparation of, 275
  use in acid-base titrations, 58–59
Phenols, response to Millon's test, 169
Phenylalanine, negative in xanthoproteic test, 169

Phosphate
  detection in nucleic acids, 255
  test for, 8, 12, 255
Phosphate buffers
  as unknowns for pH measurements, 273–274
  preparation of, 293, 296
Phosphate esters
  examples of, 112, 113, 118
  formation of, 113
  in nucleic acids, 113, 254
Phospholipids, chromatography of, 147
Phosphorus, determination of, 31–36
Photoelectric colorimeters, use of, 21
Pipet, micro
  for applying blood serum to applicator, 200
  for enzymes in cholesterol determination, 132
  for spotting in chromatography, 150
Pipetting of corrosive liquids, xi
pK$_a$
  definition of, 49
  determination of, from K$_a$ value, 49
  relation to dissociation and strength of acids, 49
Polysaccharides
  action on, by amylases, 230
  food value for man, 230
Polyunsaturated fats, 139
Ponceau S dye, as protein stain in electrophoresis, 201
Potassium acid tartrate, determination of, 59
Potassium oxalate, anticoagulant for blood, 96
Proenzymes, 221 (see also Zymogens)
Protein denaturation
  bonds affected by, 183
  effect on protein solubility, 183
  effect on protein structure and function, 183
  factors causing, 183
Protein determination
  by biuret method, 193–196
  by Lowry (Folin-Ciocalteu) method, 193
  by Kjeldahl method, 189–192
Protein precipitation
  by alkaloidal reagents, 183, 184
  by heavy metals, 183, 184
  by organic solvents, 183, 184
  by physical means, 184
  by salts, 183, 184
  by strong acids, 183, 184
Protein solubility
  effect of ammonium sulfate on, 183
  effect of pH on, 162, 183
  effect of salts on, 162, 183
Proteinases, selectivity toward peptide bonds, 221
Protein-free filtrate
  from blood plasma or serum, 95
  removal of interferences by, 95, 96
  use of zinc hydroxide-barium hydroxide to prepare, 95, 96
Proteins
  chemical elements in, 162
  color tests for, 169–174
  denaturation of, 183, 184
  detection of nitrogen in, 163
  distinguishing from fats and carbohydrates, 162, 163

electrophoresis of, 199–206
elementary composition of, 162
enzymatic digestion of, 221–226
general characteristics of, 162
general color tests for, 169–174
ignition of, 162, 163
precipitation of, 183–188
purification of, by dialysis of, 7
separation from nonprotein nitrogen
    compounds, 184, 186
solubility characteristics of, 162, 163
Proteolytic enzymes. *See* Proteinases
Provitamin A. *See* β-Carotene
Ptyalin
    action on starch and glycogen, 227
    occurrence in saliva, 227
Purine bases, test for, 255
Pyrimidine bases, test for, 255
Pyruvate, 235
Quantitative experiments
    avoiding losses in, xi
    special precautions for, xi
Raffinose, components and linkages in, 76
Reagents
    avoiding contamination of, xi
    corrosive, care in handling of, xi
    preparation of, xi, 270
Reagents, materials and equipment per
    100 students, 270–298
Records of laboratory experiments, xii
Redox dye, as hydrogen acceptor, 208
Reducing substances, interference in
    urinary glucose tests, 87, 88
Reducing sugars
    Benedict's test for, 71, 72
    definition of, 71
    determination of, in foods, 106
References, general list, 261
Resorcinol, in Seliwanoff test, 77, 78
R_f value
    calculation of, 151, 177
    definition of, 151
    of amino acids, 177, 179
    of lipids, 151, 153
RNA, color test for, 255
Saccharifying enzymes, action on starch
    and glycogen, 227
Safety in the laboratory, xi
Sakaguchi test
    chemistry of, 169
    for arginine, 170
Salicyclic acid
    in Millon's test, 170
    in preparation of aspirin, 112
    in preparation of methyl salicylate, 112
Salting-out
    of proteins, 183, 184
    of soap, 120
Saponification
    of esters, 112
    of ethyl butyrate, 112, 114
    of fats, 119–124
    general equation for, 112, 119
Saponification number
    definition of, 119
    of animal fats and vegetable oils, 119
    significance of, 119
Seliwanoff test, for ketoses, 77, 78
Serum albumin, as protein standard, 193,
    194
Serum proteins, separation by
    electrophoresis, 199–206
Sevag test, for proteins, 184

Silica gel for TLC
    activation of, 147, 148, 149
    coating glass plates with, 148–149
Smogyi-Nelson method for determination
    of blood glucose, 95–104
Soaps
    alkalinity of, 120
    hard and soft, 119
    insolubility of Ca and Mg, 120
    purification by salting-out, 120
Soda lime, equation for protein
    decomposition by, 162
Sodium bicarbonate, determination of, in
    baking soda, 59
Sodium cholate, effect on fat digestion by
    lipase, 231
Sodium hydroxide
    concentrated solutions, 268
    preparation of dilute solutions, 268
    standardization of normal solutions,
    269
Solvents, for fats, 155
Soxhlet extraction apparatus, 155, 285
Spectral range
    expressed in nanometers, 17
    of near-infrared light, 18
    of ultraviolet light, 18
    of visible light, 18
Spectronic 20 spectrophotometer
    illustration of, 19
    optical system of, 19
    use of, 21
Spectrophotometer
    illustration of, 19
    optical system of, 18, 19
    use of, 21
Standard curves for colorimetric
    determination
    of cholesterol in blood serum, 131, 132
    of glucose and lactose in foods, 107
    of glucose in blood plasma, 99
    of iron, 19–21, 33, 35
    of phosphorus, 32, 33, 35
    of proteins by biuret method, 194, 195
Standard solutions
    for acid-base titrations, 57
    preparation of, 269
Starch
    acid hydrolysis of, 72
    hydrolysis by amylases, 227–230
Starch indicator solution, for iodine
    titrations, 106, 140
Steroids, definition of, 125
Sterols
    definition of, 125
    test for, 125
Stopcock
    glass, lubrication of, 58
    Teflon, 58
Sucrose, hydrolysis of, 72, 105, 106
Sugars
    determination in foods, 105–110
    identification of unknowns, 83–86
    oxidation of, 71
Sugars in foods, 105–110
    chromatographic methods for, 105
    determination of, 105–107
    occurrence of, 105
    reactions during analysis for, 105
Sugar tests, summary of, 83–86
Sulfur
    labile, in amino acids, 169, 170
    test for, in dry ash solution, 12
    test for, in proteins, 163

Sulfuric acid
    preparation of dilute solutions, 268
    standardization of normal solutions,
    269
Sulfuric acid, concentrated
    concentration values of, 268
    corrosiveness of, 72
    dehydration of carbohydrates by, 66
    dispensing buret for, 268
    precautions in use of, 66, 129–131
Sumner units, urease activity as, 250, 297
Syrups, determination of sugars in, 106
Tashiro's indicator (Methyl red-
    methylene blue)
    color comparison flasks with, 59
    for ammonia titration, 59
    preparation of, 275
Technique. *See* Laboratory technique
TES-TAPE, for urinary glucose test, 87
Thimbles, for Soxhlet fat extraction, 155,
    156
Thin-layer chromatograms
    precoated with silica gel, 148
    visualization of, components in, 148,
    150
Thin-layer chromatography, 147–154
    advantages over paper chromatography,
    148
    mobile phase, 147
    preparation of glass plates for, 148–149
    preparation of precoated sheets for, 148
    sample spotting technique, 149
    solvent systems for, 150
    stationary phase in, 147
Titration, acid-base, 57–60
TLC. *See* Thin-layer chromatography
Tocopherols (vitamin E)
    in fat preservation, 125
    test for, 125
Total acidity
    definition of, 57
    determination of, 57–64
Trace elements
    amount in biological samples, 31
    spectrophotometric methods of analysis,.
    31
Transmittance
    mathematical relation to absorbance,
    18
    percent of, 18
    relation to incident and transmitted
    light, 18
Triacylglycerols
    enzymatic hydrolysis of, 231
    saponification of, 119
    structure of, 119
Triglycerides. *See* Triacylglycerols
Triketohydrindene hydrate. *See*
    Ninhydrin
Tris buffer, preparation of, 51, 52
Trypsin, in pancreatin powder, 221
Tryptophan
    Hopkins-Cole test for, 169, 170
    xanthoproteic test for, 169, 170
Tyrosine
    Millon's test for, 169, 170
    xanthoproteic test for, 169, 170
Ultraviolet light for visualization of
    components in chromatograms,
    148
Ultraviolet range
    in spectrophotometry, 18
    substances absorbing in, 18
Unsaturated fats, determination of iodine
    number, 139–146